AF756988

Fractional Calculus - From Theory to Applications

Edited by Christos Volos
and Jesus Manuel Muñoz-Pacheco

Published in London, United Kingdom

Fractional Calculus - From Theory to Applications
http://dx.doi.org/10.5772/intechopen.1003395
Edited by Christos Volos and Jesus Manuel Muñoz-Pacheco

Contributors
Amira M'boudi, Aslıhan Kartcı, Benaouda Hedia, Christos Volos, Mahmoud Hesaaraki, Masoud Bayrami, Nadia Allouch, Noreen Azhar, Raheela Manzoor, Saleem Iqbal, Samira Hamani, Shazia Kalsoom, Umut Engin Ayten, Yunus Emre Yokuş

© The Editor(s) and the Author(s) 2025

The rights of the editor(s) and the author(s) have been asserted in accordance with the Copyright, Designs and Patents Act 1988. All rights to the book as a whole are reserved by INTECHOPEN LIMITED. The book as a whole (compilation) cannot be reproduced, distributed or used for commercial or non-commercial purposes without INTECHOPEN LIMITED's written permission. Enquiries concerning the use of the book should be directed to INTECHOPEN LIMITED rights and permissions department (permissions@intechopen.com).

Violations are liable to prosecution under the governing Copyright Law.

Individual chapters of this publication are distributed under the terms of the Creative Commons Attribution 4.0 License which permits commercial use, distribution and reproduction of the individual chapters, provided the original author(s) and source publication are appropriately acknowledged. If so indicated, certain images may not be included under the Creative Commons license. In such cases users will need to obtain permission from the license holder to reproduce the material. More details and guidelines concerning content reuse and adaptation can be found at http://www.intechopen.com/copyright-policy.html.

Notice

Statements and opinions expressed in the chapters are these of the individual contributors and not necessarily those of the editors or publisher. No responsibility is accepted for the accuracy of information contained in the published chapters. The publisher assumes no responsibility for any damage or injury to persons or property arising out of the use of any materials, instructions, methods or ideas contained in the book.

First published in London, United Kingdom, 2025 by IntechOpen
IntechOpen is the global imprint of INTECHOPEN LIMITED, registered in England and Wales, registration number: 11086078, 167-169 Great Portland Street, London, W1W 5PF, United Kingdom

For EU product safety concerns: IN TECH d.o.o., Prolaz Marije Krucifikse Kozulić 3, 51000 Rijeka, Croatia, info@intechopen.com or visit our website at intechopen.com.

British Library Cataloguing-in-Publication Data
A catalogue record for this book is available from the British Library

Fractional Calculus - From Theory to Applications
Edited by Christos Volos and Jesus Manuel Muñoz-Pacheco
p. cm.
Print ISBN 978-1-83634-173-4
Online ISBN 978-1-83634-172-7
eBook (PDF) ISBN 978-1-83634-174-1

If disposing of this product, please recycle the paper responsibly.

We are IntechOpen, the world's leading publisher of Open Access books

Built by scientists, for scientists

7,500+
Open access books available

196,000+
International authors and editors

215M+
Downloads

156
Countries delivered to

Our authors are among the
Top 1%
most cited scientists

12.2%
Contributors from top 500 universities

CLARIVATE ANALYTICS BOOK CITATION INDEX INDEXED

WEB OF SCIENCE™

Selection of our books indexed in the Book Citation Index in Web of Science™ Core Collection (BKCI)

Interested in publishing with us?
Contact book.department@intechopen.com

Numbers displayed above are based on latest data collected.
For more information visit www.intechopen.com

Meet the editors

Dr. Christos Volos is a Full Professor in the Laboratory of Nonlinear Systems, Circuits & Complexity (LaNSCom) of the Physics Department of the Aristotle University of Thessaloniki, Greece. His research interests include chaos, nonlinear systems, mem-elements, analog and mixed-signal electronic circuits, neural networks, chaotic electronics and their applications (secure communications, cryptography, robotics), as well as chaotic synchronization and control. Dr. Volos has published over 400 papers and chapters in international journals, scientific books, and conference proceedings, in addition to editing 10 books and serving as an author on two books on the topic of nonlinear circuits and systems. Furthermore, he serves as an Editorial Board Member of 11 international journals and has been a Guest Editor of over 30 Special Issues in international journals.

Jesus Manuel Muñoz-Pacheco obtained his Ph.D. in Electronic Sciences from INAOE in 2009. He is currently a Full Professor-Researcher at BUAP. In 2018, he founded the Ph.D. program in Applied Research in Electronics, where he currently is the academic coordinator. He also established the research group "Chaos, Fractals, and Complexity" at the same university. Dr. Muñoz-Pacheco has received several recognitions for his research work, including being included in the list of the Top 2% of leading scientists globally in his area of knowledge, as awarded by Stanford University and Elsevier from 2020 to 2024. He is a Principal Fellow at the Kaabar-Wang Tech Institute in Uzbekistan, as well as serving as an External Advisor for the Research Center on Numerical Programming and Automatic Control, affiliated with the National Autonomous University of Honduras. He is also a founding member of the Mexican Society of Dynamical Systems and Complexity A. C. and has been part of the National System of Researchers (SNI) of CONAHCYT/Mexico for more than 13 consecutive years. Dr. Muñoz Pacheco has authored approximately 137 publications in scientific journals, books, conferences, and patents, which have garnered around 2246 citations - April 25 (excluding self-citations), thereby contributing to the advancement of science and technology in the areas of dynamical systems, chaos, and circuits. He is currently the Associate Editor of the *International Journal of Numerical Modeling - Electronic Networks, Devices, and Fields; Frontiers in Physics; Chaos Theory and Applications; Frontiers in Applied Mathematics and Statistics; Complexity; and Discrete Dynamics in Nature and Society.*

Contents

Preface

Fractional calculus can be considered both an old and yet a modern field of study. As it is known today, fractional calculus is a branch of mathematical physics that deals with integro-differential equations, where the integrals are typically of convolution type and involve highly singular kernels, such as power law or logarithmic functions. Although the term "fractional" is retained for historical reasons, this framework allows the computation of integrals and derivatives of any positive order. While efficient analytical and numerical methods have been developed, the field still requires careful study and attention.

But why is fractional calculus considered so important today?

Until recently, this mathematical approach was regarded as an abstract theory with little practical relevance. However, in the past decade, research activity in this area has surged, leading to applications in a wide range of scientific disciplines. Fractional calculus has proven valuable in many domains, from the physics of diffusion and transport phenomena to control theory and financial systems. More specifically, research and applications related to fractional calculus have emerged in the following fields:

- Development of analytical and numerical techniques for fractional-order calculations.
- Control and regulation of engineering systems.
- Fundamental investigations into the mechanical, electrical, and thermal properties of various materials, such as viscoelastic polymers, foams, gels, and biological tissues, both for theoretical understanding and practical applications.
- Study of wave propagation and diffusion processes, along with their experimental validation, particularly in plasma physics, such as diffusion in Tokamak systems.
- Applications in biomedical engineering and medical sciences.
- Thermal analysis of engineering systems, including braking mechanisms and machine tools.
- Image and signal processing techniques.

Therefore, this book is a collection of six chapters whose purpose is to reflect the recent development in fractional calculus, its theory, and its applications.

In more detail, the first chapter is a short introductory chapter to the history of this fascinating subject. The main milestones in the evolution of the theory of fractional calculus are mentioned, while the theoretical backgrounds, as well as the most important fields of application, are briefly mentioned.

Chapter 2 is dedicated to a new fractional derivative, which is -Riemann-Liouville and Caputo tempered χ fractional derivative, which constitutes a generalization of Riemann-Liouville and Caputo fractional derivative. Therefore, in this chapter, some properties of this new fractional derivative are presented, and the topological structure of the solution set of the fractional differential equation with mixed derivative is studied.

The next chapter presents the existence and uniqueness of solutions to the boundary value problem for fractional q-difference equations, which involves Caputo's fractional q-derivative with nonlinear integral conditions by using some standard fixed point theorems. Also, the Ulam-Hyers and Ulam-Hyers-Rassias stabilities of their solutions are studied. Furthermore, two examples are presented.

Chapter 4 briefly overviews the results of the following singular partial differential equation governed by the fractional Laplacian operator, involving singular nonlinearities and irregular data. Therefore, this chapter explores the existence and uniqueness of positive solutions to a singular fractional elliptic problem governed by the fractional Laplacian, involving singular nonlinearities and irregular data. By employing both weak and entropy solutions, the existence of solutions for a wide range of cases, including the important scenario of measuring data is studied.

Chapter 5 presents the application of fractional integral transform in the fuzzy field by presenting the solution of fuzzy fractional differential equations with the help of different integral transforms, such as the Laplace, Mellin, and Sumudu fuzzy transforms. The key features of this work are discussing different integral transforms in a fuzzy theory and naming it fuzzy integral transform with all its basic concepts.

Finally, Chapter 6 presents an in-depth analysis of pertinent publications in the rapidly growing fractional order control literature, highlighting the milestones, advantages, and disadvantages, as well as the future perspectives for the industrialization of fractional order control.

We hope this book can serve as an updated and handy reference for university professors, graduate students, laboratory researchers, physicists, and applied mathematicians interested in studying fractional calculus and its applications.

Christos Volos
Laboratory of Nonlinear Systems,
Circuits and Complexity (LanSCom),
Physics Department,
Aristotle University of Thessaloniki,
Thessaloniki, Greece

Chapter 1

Introductory Chapter: Fractional Calculus – From Theory to Applications

Christos Volos

1. Introduction

As it is known, fractional calculus is a branch of mathematical analysis that generalizes the concepts of derivation and integration to non-integer orders. Fractional calculus, although reported many centuries ago, has emerged in recent years as a powerful tool in theoretical and applied sciences. In contrast to classical calculus, which is limited to derivatives and integrals of integer-order, fractional calculus allows the processing of these operations in fractional orders, offering a powerful tool for modeling complex phenomena. This introductory chapter presents a brief review of the historical development of fractional calculus, its basic principles, and its various applications, preparing the ground for a more in-depth study in the following chapters.

The idea of fractional calculus can be traced for the first time to a letter exchange between Gottfried Wilhelm Leibniz and the Marquis de l'Hôpital in 1695, in which Leibniz expressed the idea of the existence of derivatives of non-integer orders [1]. In the following centuries, mathematicians such as Euler, Laplace, and Riemann made significant contributions to the development of the field, laying the foundations for modern fractional calculus. In particular, Leonhard Euler studied the concept of fractional factors through the gamma function, providing a mathematical base for further research [2]. Later, Pierre-Simon Laplace and Joseph Liouville expanded on these ideas, with Liouville introducing the concept of fractional integrals in the nineteenth century [3]. Bernhard Riemann refined these concepts, leading to the formulation of the Riemann-Liouville fractional derivative, which is still one of the most widely used definitions [4]. These fundamental contributions formed the basis for the modern theory of fractional calculus, which has been further enriched by the work of mathematicians such as Marcel Riesz [5], Andrey Kolmogorov [6], and, more recently, Igor Podlubny [7] and Keith Oldham [8].

Consequently, fractional calculus has evolved from a mathematical curiosity into a powerful tool that finds important applications in many scientific disciplines. Furthermore, the ability of fractional calculus to describe phenomena from systems with memory, that is, systems where the current state depends on the entire previous dynamics of the system, makes fractional calculus particularly suitable for modeling phenomena that classical calculus cannot adequately describe. For example, in viscoelastic materials, the stress at a given time is affected by the complete history of strain, which is a behavior that is naturally described by fractional derivatives [9]. Similarly, in biological systems, the evolution of populations exhibits memory phenomena, which fractional calculus can describe more accurately than classical integer-order differential equations [10]. Thus, fractional calculus is no longer just a theoretical subject but a fundamental tool with wide application in fluid dynamics, control theory, signal processing, and even financial analysis [11].

As we move into the twenty-first century, the importance of fractional calculus continues to grow due to the rapid development of computing power and the increasing complexity of the systems that we have to study. In addition, numerical methods for solving fractional differential equations have improved significantly, allowing the analysis of problems that were previously considered difficult. At the same time, the introduction of new fractional operators, such as the Atangana-Baleanu derivatives [12] and Caputo-Fabrizio derivatives [13], has expanded the capabilities of fractional calculus, providing new ways to model systems with non-local and non-singular kernels.

Overall, fractional calculus represents an innovative approach to mathematical modeling, offering new ways to understand and describe physical, biological, and engineering systems. Its historical roots, fundamental principles, and broad applications highlight its importance both as a tool for theoretical investigation and as a method for practical problem-solving. With science and engineering constantly evolving, fractional calculus is expected to play an increasingly critical role, offering new perspectives and solutions to the challenges of the future.

2. Fundamental concepts and definitions

At the heart of fractional calculus lies the generalization of the derivative and integral operators. The most commonly used definitions include the Riemann-Liouville, Caputo, and Grünwald-Letnikov fractional derivatives. Each of these definitions has its own advantages and is suited to different types of problems.

The Riemann-Liouville fractional derivative is defined as [14]:

$$D^{\alpha} f(t) = \frac{1}{\Gamma(n-\alpha)} \frac{d^n}{dt^n} \int_a^t \frac{f(\tau)}{(t-\tau)^{\alpha-n+1}} d\tau \tag{1}$$

where α is the fractional-order, n is the smallest integer greater than α, $\Gamma(\cdot)$ is the gamma function, and $\frac{d^n}{dt^n}$ represents the standard integer-order derivative. This definition is particularly useful in theoretical studies due to its strong mathematical foundation.

On the other hand, the Caputo fractional derivative is defined as [15]:

$$D^{\alpha} f(t) = \frac{1}{\Gamma(n-\alpha)} \int_a^t \frac{f^{(n)}(\tau)}{(t-\tau)^{\alpha-n+1}} d\tau \tag{2}$$

where $f^{(n)}(t)$ is the *integer-order derivative* of $f(t)$. This definition is often preferred in applied sciences because it allows for the inclusion of initial conditions in a manner similar to classical differential equations.

Furthermore, the Grünwald-Letnikov fractional derivative provides a discrete approximation and is defined as [16]:

$$D^{\alpha} f(t) = \lim_{h \to 0} \frac{1}{h^{\alpha}} \sum_{\kappa=0}^{\infty} (-1)^{k} \binom{a}{k} f(t - kh) \tag{3}$$

where h is a small step size and $\binom{a}{k}$ is the *generalized binomial coefficient*, defined as:

$$\binom{a}{k} = \frac{\Gamma(a+1)}{\Gamma(k+1)\Gamma(a-k+1)} \tag{4}$$

This definition is particularly useful in numerical simulations and computational applications.

3. Applications of fractional calculus

As it has been mentioned, fractional calculus has found applications in a wide range of fields, demonstrating its versatility and power. Below, a few key areas where fractional calculus has made significant contributions are highlighted.

- *Fluid dynamics*

In the field of fluid dynamics, fractional calculus is a powerful tool for modeling non-Newtonian fluids with complex rheological properties. Models based on fractional derivatives represent more accurately the viscoelastic behavior of these fluids, providing information about the system that cannot be extracted with classical approaches [9].

- *Control theory*

Fractional-order controllers, such as Fractional-Order Proportional-Integral-Derivative (FOPID), have many advantages in applications where control of systems with large memory is required [17]. Compared to classical integer-order controllers, the fractional-order controllers offer increased flexibility and improved robustness against uncertainties and systems' variations.

- *Signal processing*

In signal analysis and processing, fractional calculus is used to design fractional-order filters and to study signals with large-scale dependence properties [18]. Therefore, these tools are particularly useful for analyzing signals that exhibit long-memory characteristics, such as biological and financial data.

- *Finance*

In the field of financial mathematics, fractional calculus is used to model market dynamics and option pricing [11]. A typical example is fractional Brownian motion, which offers a more accurate description of market behavior by taking into account the long-term dependencies that appear in financial time series.

4. Conclusion

Fractional calculus represents a significant advancement in mathematical analysis, providing an expanded and more flexible framework for modeling complex systems. The ability of fractional calculus to describe phenomena with memory and non-local interactions makes it an indispensable tool in various scientific and engineering disciplines. As research in this field continues, the applications of fractional calculus are expected to expand even further, enhancing both its theoretical importance and its usefulness in practical problems.

Author details

Christos Volos
Laboratory of Nonlinear Systems, Circuits and Complexity Department of Physics, Aristotle University of Thessaloniki, Greece

*Address all correspondence to: volos@physics.auth.gr

IntechOpen

© 2025 The Author(s). Licensee IntechOpen. This chapter is distributed under the terms of the Creative Commons Attribution License (http://creativecommons.org/licenses/by/4.0), which permits unrestricted use, distribution, and reproduction in any medium, provided the original work is properly cited.

References

[1] Arthur RT, Rabouin D. Leibniz on the Foundations of the Differential Calculus. Switzerland: Springer Nature; 2025

[2] Gautschi W. Leonhard Euler: His life, the man, and his works. SIAM Review. 2008;**50**(1):3-33

[3] Ross B. The development of fractional calculus 1695-1900. Historia Mathematica. 1977;**4**(1):75-89

[4] Nonnenmacher TF, Metzler R. On the Riemann-Liouville fractional calculus and some recent applications. Fractals. 1995;**3**(03):557-566

[5] Tarasov VE. General fractional calculus in multi-dimensional space: Riesz form. Mathematics. 2023;**11**(7):1651

[6] Tenreiro Machado JA. Fractional coins and fractional derivatives. Abstract and Applied Analysis. 2013;**2013**(1):205097

[7] Podlubny I. Fractional Differential Equations. San Diego: Academic Press; 1999

[8] Oldham KB. Fractional differential equations in electrochemistry. Advances in Engineering Software. 2010;**41**(1):9-12

[9] Adolfsson K, Enelund M, Olsson P. On the fractional order model of viscoelasticity. Mechanics of Time Dependent Materials. 2005;**9**:15-34

[10] Nisar KS, Farman M, Abdel-Aty M, Ravichandran C. A review of fractional order epidemic models for life sciences problems: Past, present and future. Alexandria Engineering Journal. 2024;**95**:283-305

[11] Kilbas AA, Srivastava HM, Trujillo JJ. Theory and Applications of Fractional Differential Equations. Amsterdam: Elsevier; 2006

[12] Syam MI, Al-Refai M. Fractional differential equations with Atangana–Baleanu fractional derivative: Analysis and applications. Chaos, Solitons & Fractals:X. 2019;**2**:100013

[13] Al-Refai M, Pal K. New aspects of Caputo–Fabrizio fractional derivative. Progress in Fractional Differentiation & Applications. 2019;**5**(2):157-166

[14] Li C, Qian D, Chen Y. On Riemann-Liouville and Caputo derivatives. Discrete Dynamics in Nature and Society. 2011;**2011**(1):562494

[15] Farid G, Latif N, Anwar M, Imran A, Ozair M, Nawaz M. On applications of Caputo k-fractional derivatives. Adv. Difference Equ. 2019;**2019**:1-16

[16] Scherer R, Kalla SL, Tang Y, Huang J. The Grünwald–Letnikov method for fractional differential equations. Computers & Mathematics with Applications. 2011;**62**(3):902-917

[17] Padula F, Visioli A. Optimal tuning rules for proportional-integral-derivative and fractional-order proportional-integral-derivative controllers for integral and unstable processes. IET Control Theory and Applications. 2012;**6**(6):776-786

[18] Gonzalez EA, Petráš I. Advances in fractional calculus: Control and signal processing applications. In: Proceedings of the 2015 16th international Carpathian Control Conference (ICCC). Szilvasvarad, Hungary: IEEE Xplore; 2015. pp. 147-152

Chapter 2

An Existence Results for Tempered Fractional Differential Equation with ϕ-Fractional Derivative

Amira M'boudi and Benaouda Hedia

Abstract

We define firstly a new fractional derivative, which is Riemann-Liuoville and Caputo tempered χ fractional derivative, it constitutes a generalization of Riemann-Liouville and Caputo fractional derivative, and we give also some properties that will be a key tool during this paper; secondly, we study the topological structure of the solution set of fractional differential equation with mixed derivative. We use the technique of condensing operator combined with Kuratowski measure of non-compactness to prove that the solution set is nonempty and compact. Finally, we give an example to illustrate the feasibility of our main result.

Keywords: ϕ-fractional double derivative, fixed point theorem, fractional differential equations, measure of non-compactness, Meir-Keeler condensing operator

1. Introduction

Fractional differential equations have become more important in the recent years in some mathematical models of real processes and phenomena studied in different area. We refer to the papers cited below for example [1–7]. Consequently, the fractional calculus and its applications in several fields of science and engineering have received much attention and have been developed very rapidly.

The monographs [8–12] contain analysis tools basic and foundation of fractional calculus.

In the last few years, famous papers treated the study of a special fractional differential equations with initial, boundary, and integral boundary conditions by using the fixed point theory combined with measure of non-compactness involving ϕ and tempered fractional derivative [13–18].

We consider, in this work, the following fractional differential equation We shall study the existence solution set and its topological structure of abstract fractional equation in a weight Banach space.

$$(\mathfrak{P}) \quad \begin{cases} {}^{rl}_{\lambda}{}^{T}\mathcal{D}^{\xi,\chi}_{0^+}({}^{C}_{\mu}{}^{T}\mathcal{D}^{\zeta,\chi})y(r) = f(\wp, y(\wp), {}^{C}D^{\zeta,\chi}y(\wp)), \quad \in (0, L], \\ {}^{T}_{\lambda}\mathfrak{J}^{(1-\xi),\chi}_{0^+}({}^{C}_{\mu}{}^{T}\mathcal{D}^{\zeta,\chi})y(0^+) = \rho_1, \\ y(0) = \rho_2, \end{cases}$$

IntechOpen

where ${}^{rlT}_{\lambda}\mathcal{D}^{\xi,\chi}_{0^+}$ (resp. ${}^{CT}_{\mu}\mathcal{D}^{\zeta,\chi}$) denotes the left-sided χ-Riemann-Liouville (resp. χ-Caputo) tempered fractional derivative of order ξ, ζ with $0 < \xi < 1, 0 < \zeta < 1$ and $\xi + \zeta > 1$ and $\lambda > 0, \mu > 0$. The operator $\mathfrak{I}^{(1-\xi),\lambda,\chi}_{0^+}$ denotes the left-sided χ-Riemann-Liouville tempered fractional integral, $\mathfrak{E}$ is a Banach space with the norm $\|.\|, \rho_1, \rho_2 \in \mathfrak{E}, f : (0, L] \times \mathfrak{E} \times \mathfrak{E} \to \mathfrak{E}$, a function satisfying some specified conditions (see, Section 3), and $\chi \in C^1([0, L], \mathbb{R}^+)$ satisfies $\chi'(\wp) > 0$, for all $r \in [0, L]$.

This article is organized as follows. In Section 2, we introduce some preliminaries results needed in the following sections. In Section 3, we prove existence results and its topological structure for the problem $(\mathfrak{P})$ under some imposed hypothesis. In Section 4, we present an example in which we illustrate the feasibility of the problem considered.

2. Preliminaries

In this section, we state some definitions and results that will be used in this work.

Let $C([0, L], \mathfrak{E})$ be the space of E-valued continuous functions on $[0, L]$ endowed with the uniform norm topology

$$\|x\|_\infty = \sup\{\|x(\wp)\|, \quad \wp \in [0, L]\}.$$

Let $C_{\xi,\chi}([0, L], \mathfrak{E})$ and $C^{\zeta}_{\xi,\chi}([0, L], \mathfrak{E})$ two Banach spaces of functions defined as follows:

$$C_{\xi,\chi}([0, L], \mathfrak{E}) = \{\tau \in C((0, L], \mathfrak{E}) : \quad \lim_{\wp \to 0^+} (\tau(r) - \tau(0))^{1-\xi}\tau(\wp) \ \text{ exists and finite}\},$$

equipped with the norm

$$\|\tau\|_{C_{\xi,\chi}} = \sup_{\wp \in [0, L]} (\chi(\wp) - \chi(0))^{1-\xi}\|\tau(\wp)\|,$$

and

$$C^{\zeta}_{\xi,\chi}([0, L], \mathfrak{E}) = \{\tau : [0, L] \to \mathfrak{E} : \tau \in C([0, L], \mathfrak{E}) \text{ and } {}^{c}\mathcal{D}^{\zeta,\chi}\tau \in C_{\xi,\chi}([0, L], \mathfrak{E})\},$$

with the norm

$$\|\tau\|_{C^{\zeta}_{\xi,\chi}} = \sup_{\wp \in [0, L]} \|\tau(\wp)\| + \sup_{\wp \in [0, L]} (\chi(\wp) - \chi(0))^{1-\xi}\|{}^{c}\mathcal{D}^{\zeta,\chi}\tau(\wp)\|.$$

2.1 Fractional integral

We begin with some definitions from the theory of fractional calculus. For all $\eta > -1$ and $s, r \in [0, L]$ with $r \geq s$, we pose $\psi_\eta(r, s) = (\chi(r) - \chi(s))^\eta$.

Definition 2.1. [13]. (χ-Fractional Integral).

Let $\delta \in E^p(\varrho_1, \varrho_2)$ and $[\varrho_1, \varrho_2]$ be a finite or infinite interval of $\mathbb{R}, \chi(d) > 0$ be an increasing function on $(\varrho_1, \varrho_2]$ and $\chi'(d) > 0$ be continuous on $[\varrho_1, \varrho_2]$ and $\xi > 0$. The fractional integral operators of a function δ of order ξ are given by

$$\mathfrak{I}^{\xi,\chi}_{\varrho_1^+}\delta(r) = \int_{\varrho_1}^{r} \chi'(s)\psi_{\xi}(r,s)\delta(s)ds,$$

$$\mathfrak{I}^{\xi,\chi}_{\varrho_2^-}\delta(r) = \int_{r}^{\varrho_2} \chi'(s)\psi_{\xi}(r,s)\delta(s)ds.$$

with $\psi_{\xi}(r,s) = \frac{(\chi(r)-\chi(s))^{\xi-1}}{\Gamma(\xi)}$.

Definition 2.2. [8]. (χ-Tempered Fractional Integral).

Let $\delta \in E^p(\varrho_1, \varrho_2)$ and $[\varrho_1, \varrho_2]$ be a finite or infinite interval of $\mathbb{R}, \chi(d) > 0$ be an increasing function on $(\varrho_1, \varrho_2]$ and $\chi'(d) > 0$ be continuous on $[\varrho_1, \varrho_2]$ and $\xi > 0$. The tempered fractional integral operators of a function δ of order ξ are given by

$${}^{T}_{\lambda}\mathfrak{I}^{\xi,\chi}_{\varrho_1^+}\delta(r) - \int_{\varrho_1}^{r} \chi'(s)\psi_{\xi-1}(r,s)e^{-\lambda(\chi(r)-\chi(s))}\delta(s)ds,$$

$${}^{T}_{\lambda}\mathfrak{I}^{\xi,\chi}_{\varrho_2^-}\delta(r) = \int_{r}^{\varrho_2} \chi'(s)\psi_{\xi-1}(r,s)e^{-\lambda(\chi(r)-\chi(s))}\delta(s)ds.$$

with $\psi_{\xi}(r,s) = \frac{(\chi(r)-\chi(s))^{\xi}}{\Gamma(\xi)}$.

By embodying Definition (2.1) and Definition (2.2), we can now present the ensuing definition of an extensive fractional integral that embraces both integrals as specific instances.

2.2 Fractional derivatives

Definition 2.3. (χ – *Riemann Liouville fractional Derivative*). [13]

Let $\xi > 0$ a real number and $\nabla = [\varrho_1, \varrho_2]$ an interval such that $-\infty \leq \varrho_1 < \varrho_2 \leq +\infty, \xi > 0,\ n = [\xi] + 1.$ and $\delta, \chi \in C^1([\varrho_1, \varrho_2], \mathbb{R})$ two functions such that χ is increasing and $\chi(r) \neq 0$, for all $r \in \nabla$. The χ Riemann-Liouville fractional derivatives $\mathfrak{D}^{\xi,\chi}_{\varrho_1^+}$ and $\mathfrak{D}^{\xi,\chi}_{\varrho_2^+}$ of a function δ with respect to χ of order ξ are defined by

$$\begin{aligned} {}^{rl}D^{\xi,\chi}_{\varrho_1^+}\delta(r) &= \left(\frac{1}{\chi'(r)}\frac{d}{dr}\right)^n I^{n-\xi,\chi}_{\varrho_1^+}\delta(r) \\ &= \frac{1}{\Gamma(n-\xi)}\left(\frac{1}{\chi'(r)}\frac{d}{dr}\right)^n \int_{\varrho_1}^{r} \chi'(s)[\chi(r)-\chi(s)]^{n-\xi-1}h(t)dt. \end{aligned}$$

$$\begin{aligned} D^{\xi,\chi}_{\varrho_2^-}\delta(r) &= \left(-\frac{1}{\chi'(r)}\frac{d}{r}\right)^n I^{n-\xi,\chi}_{\varrho_2^-}\delta(r) \\ &= \frac{1}{\Gamma(n-\xi)}\left(-\frac{1}{\chi'(r)}\frac{d}{dr}\right)^n \int_{r}^{\varrho_2} \chi'(s)[\chi(s)-\chi(r)]^{n-\xi-1}\delta(s)ds. \end{aligned}$$

Definition 2.4. [13] Let $n - 1 < \xi < n$ with $n \in \mathbb{N}^*$, $\nabla = [\varrho_1, \varrho_2]$ be such that $-\infty \leq \varrho_1 < \varrho 2 \leq \infty$ and $\delta, \chi \in C^n([\varrho_1, \varrho_2], \mathbb{R})$ be two functions such that χ is nondecreasing and $\chi'(r) \neq 0$ for all $t \in [\varrho_1, \varrho_2]$. The left side χ-Caputo fractional derivative (respectively the right-sided one) ${}^{C}\mathbb{D}^{\xi,\chi}_{\varrho_1^+}(.)$ and ${}^{C}\mathbb{D}^{\xi,\chi}_{\varrho_2^-}(.)$ of function δ of order ξ are defined by

$$
\begin{aligned}
{}^{C}\mathbb{D}_{\varrho_1^+}^{\xi,\chi}\delta(r) &= I_{\varrho_1^+}^{(n-\xi),\chi}\left(\frac{1}{\chi'(r)}\frac{d}{dr}\right)^n \delta(r) \\
&= I_{\varrho_1^+}^{1-\xi,\chi} D_{\varrho_1^+}^{\gamma,\chi}\delta(r), \\
{}^{C}\mathbb{D}_{\varrho_2^-}^{\xi,\chi}\delta(r) &= I_{\varrho_2^-}^{(n-\xi),\chi}\left(-\frac{1}{\chi'(t)}\frac{d}{dr}\right)^n \delta(r) \\
&= I_{\varrho_2^-}^{1-\xi,\psi}(-1)^n D_{\varrho_2^-}^{1,\chi}\delta(r),
\end{aligned}
$$

$I_{\varrho_1^+}^{1-\xi,\chi}(.), D_{\varrho_1^+}^{1,\chi}(.), I_{\varrho_2^-}^{1-\xi,\chi}, D_{\varrho_2^-}^{1,\chi}(.)$ are left-sided and right-sided χ-Riemann-Liouville fractional integral and derivative, respectively.

We need the notations,

$$
\delta_{\chi^+}^{[n]}\delta(r) := \left(\frac{1}{\chi'(r)}\frac{d}{dr}\right)^n \delta(r) \text{ and } \delta_{\chi^-}^{[n]}\delta(r) := \left(-\frac{1}{\chi'(r)}\frac{d}{dr}\right)^n \delta(r).
$$

Definition 2.5. (Tempered χ – *Riemann Liouville fractional derivative*) [13].
Let $\xi > 0$ a real number and $\nabla = [\varrho_1, \varrho_2]$ an interval such that $-\infty \le \varrho_1 < \varrho_2 \le +\infty$, $\xi > 0$, $n = [\xi] + 1$. and $\delta, \chi \in C^1([\varrho_1, \varrho_2], \mathbb{R})$ two functions such that χ is increasing and $\chi(r) \neq 0$, for all $r \in \nabla$. The tempered χ Riemann-Liouville fractional derivatives ${}^{rl}_{\lambda}{}^{T}\mathfrak{D}_{\varrho_1^+}^{\xi,\chi}$ and ${}^{rl}_{\lambda}{}^{T}\mathfrak{D}_{\varrho_2^+}^{\xi,\chi}$ of a function δ with respect to χ of order ξ are defined by

$$
\begin{aligned}
{}^{rl}_{\lambda}{}^{T}D_{\varrho_1^+}^{\xi,\chi}\delta(r) &= \left(\frac{1}{\chi'(r)}\frac{d}{dr} + \lambda\right)^n I_{\varrho_1^+}^{n-\xi,\chi}\delta(r) \\
&= e^{-\lambda\chi(r)rl}D_{\varrho_1^+}^{\xi,\chi}\left(e^{\lambda\chi(r)}\delta(r)\right) \\
{}^{rl}_{\lambda}{}^{T}D_{\varrho_2^-}^{\xi,\chi}\delta(r) &= \left(-\frac{1}{\chi'(r)}\frac{d}{dr} + \lambda\right)^n I_{\varrho_2^-}^{n-\xi,\chi}\delta(r) \\
&= e^{\lambda\chi(r)}D_{\varrho_1^+}^{\xi,\chi}\left(e^{-\lambda\chi(r)}\delta(r)\right).
\end{aligned}
$$

Definition 2.6. [13] Let $n - 1 < \xi < n$ with $n \in \mathbb{N}^*$, $\nabla = [\varrho_1, \varrho_2]$ be such that $-\infty \le \varrho_1 < \varrho_2 \le \infty$ and $\delta, \chi \in C^n([\varrho_1, \varrho_2], \mathbb{R})$ be two functions such that χ is increasing and $\chi'(r) \neq 0$ for all $r \in [\varrho_1, \varrho_2]$. The tempered χ-Caputo fractional derivative (left-sided and right-sided) ${}^{C}\mathbb{D}_{\varrho_1^+}^{\xi,\chi}(.)$ and ${}^{C}\mathbb{D}_{\varrho_2^-}^{\xi,\chi}(.)$ of function δ of order ξ are defined by

$$
\begin{aligned}
{}^{CT}_{\lambda}\mathbb{D}_{\varrho_1^+}^{\xi,\chi}\delta(r) &= I_{\varrho_1^+}^{(n-\xi),\chi}\left(\frac{1}{\chi'(r)}\frac{d}{dr} + \lambda\right)^n \delta(r) \\
&= e^{-\lambda\chi(r)C}\mathbb{D}_{\varrho_1^+}^{\xi,\chi}\left(e^{\lambda\chi(r)}\delta(r)\right)
\end{aligned}
$$

and

$$
\begin{aligned}
{}^{CT}_{\lambda}\mathbb{D}_{\varrho_2^+}^{\xi,\chi}\delta(r) &= I_{\varrho_2^+}^{(n-\xi),\chi}\left(-\frac{1}{\chi'(r)}\frac{d}{dr} + \lambda\right)^n \delta(r) \\
&= e^{\lambda\chi(r)C}\mathbb{D}_{\varrho_2^+}^{\xi,\chi}\left(e^{-\lambda\chi(r)}\delta(r)\right)
\end{aligned}
$$

We need the following lemma,

Lemma 2.1. Let $n-1<\xi<n$ and $\delta:[\varrho_1,\varrho_2]\to\mathbb{R}$. Then, the following holds:

1. If $\delta\in C[\varrho_1,\varrho_2]$, then ${}^{C}\mathcal{D}_a^{\xi,\chi}\mathfrak{I}_a^{\xi,\chi}\delta(r)=\delta(r)$,

2. If $\delta\in C^n[\varrho_1,\varrho_2]$, then $\mathfrak{I}_{\varrho1}^{\xi,\chi}\,{}^{C}\mathcal{D}_{\varrho1}^{\xi,\chi}\delta(r)=\delta(r)-\sum_{k=0}^{n-1}c_k[\chi(r)-\chi(\varrho_1)]^k.$

where $c_k=\frac{\delta_\chi^{[k]}(\varrho_1)}{k!}$.

Lemma 2.2. (Composition properties). If $\delta\in C^j_{\theta,\chi}[\varrho_1,\varrho_2],\ j-1<\xi<j\lambda\in\mathbb{R}$, where $j\in\mathbb{N}$ then

1. ${}_{\lambda}^{C\,T}\mathcal{D}_a^{\xi,\chi}\,{}_{\lambda}^{T}\mathfrak{I}_{\varrho1}^{\xi,\chi}\delta(r)=\delta(r)$,

2. $${}_{\lambda}^{T}\mathfrak{I}_{\varrho_1^+}^{\xi,\chi}\,{}_{\lambda}^{CT}\mathcal{D}_{\varrho_1^+}^{\xi,\chi}\delta(r)=\delta(r)-e^{-\lambda(\chi(r)-\chi(\varrho_1))}\sum_{i=1}^{j}\frac{(\chi(r)-\chi(\varrho_1))^{\xi-i}}{\Gamma(\xi-i+1)}\left\{\delta_\psi^{j-i}\mathfrak{I}_{\varrho_1^+}^{j-\xi,\chi}\delta(\varrho_1)\right\}$$
$$=\delta(r)-e^{-\lambda(\chi(r)-\chi(\varrho_1))}\sum_{i=1}^{j}c_i\psi_{\xi-i}(r,\varrho_1)$$

where $c_i=\left\{\delta_\psi^{j-i}\mathfrak{I}_{\varrho_1^+}^{j-\xi,\chi}\delta(\varrho_1^+)\right\}$,

in particular if $j=1$ we have

$${}_{\lambda}^{T}\mathfrak{I}_{\varrho1^+}^{\xi,\chi}\,{}_{\lambda}^{C\,T}\mathcal{D}_{\varrho1^+}^{\xi,\chi}\delta(r)=\delta(r)-ce^{-\lambda(\chi(t)-\chi(a))}\psi_\xi(r,\varrho_1).$$

Lemma 2.3. [8, 19] Let $\xi,\zeta\in\mathbb{R}_+^*$. Then, we have:

$${}_{\lambda}^{T}\mathfrak{I}_{\varrho_1^+}^{\xi,\chi}e^{-\lambda(\chi(r)-\chi(\varrho_1))}\psi_{\zeta-1}(r,\varrho_1)=e^{-\lambda(\chi(r)-\chi(\varrho_1))}\frac{\Gamma(\zeta)}{\Gamma(\xi+\zeta)}\psi_{\xi+\zeta-1}(r,\varrho_1),$$

and

$${}_{\lambda}^{T}\mathfrak{I}_{\varrho_2^-}^{\xi,\chi}e^{-\lambda(\chi(\varrho_2)-\chi(r))}\psi_{\zeta-1}(\varrho_2,r)=e^{-\lambda(\chi(\varrho_2)-\chi(r))}\frac{\Gamma(\zeta)}{\Gamma(\xi+\zeta)}\psi_{\xi+\zeta-1}(\varrho_2,r),$$

Lemma 2.4. [8, 19] Let $\xi,\zeta\in\mathbb{R}_+^*$. Then, we have: If $0<\xi<1$, then

$${}_{\lambda}^{rl\ T}\mathcal{D}_{0^+}^{\xi,\chi}\psi_{\xi-1}(r,\varrho_1)=0,$$

If $\zeta>1$, then

$${}_{\lambda}^{CT}\mathcal{D}_{0^+}^{\xi,\chi}\psi_{\zeta-1}(r,\varrho_1)=e^{-\lambda((\chi(r)-\chi(\varrho_1))}\frac{\Gamma(\zeta)}{\Gamma(\zeta-\xi)}\psi_{\zeta-\xi-1}(r,\varrho_1).$$

We now define the Kuratowski measure of noncompactness.

Definition 2.7. [20–22]. Let M be a metric space and X be a subset of M. The Kuratowski measure of noncompactness $\nu(X)$ of the set X is defined as

$$\nu(X):=\inf\{d>0\mid \text{there exist finitely many sets of diameter at most } d \text{ which cover } X\}.$$

We use ν and ν_C to denote the Kuratowski noncompactness measure of bounded sets in the spaces E and $C(J, E)$, respectively.

Let V a bounded set of $\mathfrak{E}$. The Kuratowski measure of non-compactness ν of the subset V is defined as follows:

$$\nu(V) = \inf\{e > 0 : \Omega \text{ admits a finite cover by sets of diameter} \leq e\}.$$

Lemma 2.5. [20, 23] Let V, W bounded sets. The following properties hold:

(i_1) $\nu(V) = 0$ if and only if V is relatively compact,

(i_2) $\nu(V) = \nu(\overline{V})$, where $\overline{V}$ denotes the closure of V,

(i_3) $\nu(V + W) \leq \nu(A) + \nu(B)$,

(i_4) $V \subset W$ implies $\nu(V) \leq \nu(W)$,

(i_5) $\nu(\rho.V) = \|\rho\|.\nu(V)$ for all $\rho \in \mathfrak{E}$,

(i_6) $\nu(\{\rho\} \cup V) = \nu(V)$ for all $\rho \in \mathfrak{E}$,

(i_7) $\nu(V) = \nu(Conv(V))$, where $Conv(V)$ is the smallest convex that contains V.

Lemma 2.6. [20–22] Let (V) a bounded set of $\mathfrak{E}$ and $\varepsilon > 0$. Then, there is a sequence $\{\varsigma_n\}_{n \in \mathbb{N}} \subset V$, such that

$$\nu(D) \leq 2\nu(\{\varsigma_n, n \in \mathbb{N}\}) + \varepsilon.$$

Lemma 2.7. [20–22] If $V \subset C(J, E)$ is a bounded and equi-continuous set, then

i) the function $t \to \nu(V(t))$ is continuous on J and

$$\nu_c(V) = \sup_{1 \leq t \leq T} \nu(V(t)),$$

ii)
$$\nu\left(\int_1^T x(s)ds : \quad x \in V\right) \leq \int_1^T \nu(V(s))ds,$$

where $V(s) = \{x(s) : \ x \in V\}, s \in J.$

The following fixed point theorem due Meir-Keeler plays a key role in the proof of our main results.

Definition 2.8. [21, 22] Let ν be an arbitrary measure of non-compactness on E and V be a nonempty subset of $\mathfrak{E}$. Let $\aleph$ be an operator from V to V. $\aleph$ is said Meir-Keeler condensing operator if

$$\forall \epsilon > 0, \ \exists \kappa(\epsilon) > 0, \forall \Omega \subset V \ \ \Omega \ \text{ bounded } : \epsilon \leq \nu(\Omega) < \epsilon + \kappa \Rightarrow \nu(\aleph\Omega) < \epsilon.$$

Theorem 1.1 [21, 22] Let ν be an arbitrary measure of non-compactness on $\mathfrak{E}$ and V a closed, bounded and convex subset of $\mathfrak{E}$. Let $\aleph$ be an operator from V to V, allow that $\aleph$ is a Meir-Keeler continuous and condensing map, then the set $\{\tau \in V : \aleph(\tau) = \tau\}$ is nonempty and compact.

Lemma 2.8. Let $\kappa \in C([0, L], \mathfrak{E})$ and $y \in C^{\zeta}_{\xi,\chi}([0, L], \mathfrak{E})$, then y satisfies the problem

$$ {}^{rl\ T}_{\lambda}\mathcal{D}^{\xi,\chi}_{0^+}\left({}^{C\ T}_{\mu}\mathcal{D}^{\zeta,\chi}\right)y(\wp) = \kappa(\wp), \quad \wp \in (0, L], \tag{1}$$

$$ {}^{T}_{\lambda}\mathfrak{I}^{(1-\xi),\chi}_{0^+}\left({}^{C\ T}_{\mu}\mathcal{D}^{\zeta,\chi}\right)y(0^+) == \rho_1, \tag{2}$$

$$ y(0) = \rho_2, \tag{3}$$

if y is solution of the following integral equation,

$$\begin{aligned} y(\wp) = \ & \rho_2 + \rho_1 e^{-(\lambda+\mu)(\chi(\wp)-\chi(0))} \frac{\psi_{\zeta+\xi-1}(\wp,0)}{\Gamma(\zeta+\xi)} \\ & + \frac{1}{\Gamma(\zeta)\Gamma(\xi)} \int_0^{\wp}\int_0^{s} \chi'(s)\chi'(\tau)\psi_{\zeta-1}(\wp,s)\psi_{\xi-1}(s,\tau)e^{-\lambda[\chi(\wp)-\chi(s)]}e^{-\lambda[\chi(s)-\chi(\tau)]}\kappa(\tau)d\tau ds, \end{aligned} \tag{4}$$

Assume y satisfies (1)-(3). By applying ${}_{\lambda}^{T}\mathfrak{I}_{0^+}^{\xi,\chi}$ to both sides of (1) and taking in account Lemma 2.2 and 2.3 one has

$$ {}_{\mu}^{CT}\mathcal{D}^{\zeta,\chi}y(\wp) = \frac{c_1 e^{-\lambda\chi(\wp)}\psi_{\xi-1}(\wp,0)}{\Gamma(\xi)} + \frac{1}{\Gamma(\xi)}\int_0^{\wp} \chi'(s)\psi_{\xi-1}(\wp,s)e^{-\lambda[\chi(\wp)-\chi(s)]}y(s)ds, \tag{5}$$

applying ${}_{\lambda}^{T}\mathfrak{I}_{0^+}^{1-\xi,\chi}$ to both sides of (5) and utilizing Lemma 2.3, we get

$$ {}_{\lambda}^{T}\mathfrak{I}_{0^+}^{1-\xi,\chi}\left({}_{\mu}^{CT}\mathcal{D}^{\zeta,\chi}y(\wp)\right) = c_1 e^{-\lambda\chi(\wp)} + {}_{\lambda}^{T}\mathfrak{I}_{0^+}^{1,\chi}\gamma(\wp)$$

By taking $\wp$ tends to 0, it follows then $c_1 = \rho_1 e^{\lambda\chi(0)}$. Hence

$$ {}_{\mu}^{CT}\mathcal{D}^{\zeta,\chi}y(\wp) = \frac{\rho_1 e^{-\lambda\chi(\wp)}\psi_{\xi-1}(\wp,0)}{\Gamma(\xi)} + \frac{1}{\Gamma(\xi)}\int_0^{\wp} \chi'(s)\psi_{\xi-1}(\wp,s)e^{-\lambda[\chi(\wp)-\chi(s)]}\kappa(s)ds, \tag{6}$$

we apply ${}_{\mu}^{T}\mathfrak{I}_{0^+}^{\zeta,\chi}$ to both sides of (6) one has

$$\begin{aligned} y(\wp) = \ & c_0 + \rho_1 e^{-(\lambda+\mu)(\chi(\wp)-\chi(0))} \frac{\psi_{\zeta+\xi-1}(\wp,0)}{\Gamma(\zeta+\xi)} \\ & + \frac{1}{\Gamma(\zeta)\Gamma(\xi)} \int_0^{\wp}\int_0^{s} \chi'(s)\chi'(\tau)\psi_{\zeta-1}(\wp,s)\psi_{\xi-1}(s,\tau)e^{-\lambda[\chi(\wp)-\chi(s)]}e^{-\lambda[\chi(s)-\chi(\tau)]}\kappa(\tau)d\tau ds, \end{aligned}$$

condition (1) yields $c_0 = \rho_2$. Consequently (1) is obtained.

Conversely, let $y \in C_{\xi,\chi}^{\zeta}([0,L],\mathfrak{E})$ be a solution of (4), we have $y(0) = \rho_2$. Next, by applying ${}_{\mu}^{CT}\mathcal{D}^{\xi,\chi}$ to both sides of (4) and taking into accounts 2.3 and 2.4, it follows then (6).

Applying ${}_{\mu}^{rl}\ {}^{T}\mathcal{D}^{\xi,\chi}$ to both sides of (6) and bearing in the mind Lemmas 2.3 and 2.4, we get (1), again by arguing as above by applying ${}_{\lambda}^{T}\mathfrak{I}_{0^+}^{(1-\xi),\chi}$ to both sides of (6) and utilizing Lemma 2.2

$$ {}_{\lambda}^{T}\mathfrak{I}_{0^+}^{(1-\xi),\chi}\left({}_{\mu}^{CT}\mathcal{D}^{\zeta,\chi}\right)y(\wp) = \rho_1 + {}_{\lambda}^{T}\mathfrak{I}_{0^+}^{1,\chi}\gamma(\wp)$$

By taking $\wp$ tends to 0, we get (2).

3. Body of the manuscript

We first put the following hypotheses:

$(\mathbf{H}_1)$ The function $f : (0,L] \times \mathfrak{E} \times \mathfrak{E} \to \mathfrak{E}$ is continuous and for all $x, y, u, v \in \mathfrak{E}$ and $r \in (0,L]$:

$$\| f(r,x,y) - f(r,u,v), \| \leq A\|x-u\| + \psi_{1-\xi}(r,0)B\|y-v,\|,$$

where $A,\ B \in \mathbb{R}^+$.

$(\mathbf{H}_2)$ There exists two functions $\imath, \jmath \in C([0,L],\mathbb{R}^+)$ such that for each nonempty, bounded set $\Omega \subset C^{\zeta}_{\xi,\chi}([0,L],\mathfrak{E})$,

$$\vartheta(f(r,\Omega(r),{}^{C}\mathcal{D}^{\zeta,\chi}\Omega(r))) \leq \imath(r)\vartheta(\Omega(r)) + \psi_{1-\xi}(r,0)\jmath(r)\vartheta({}^{C}\mathcal{D}^{\zeta,\chi}\Omega(r)), \quad \text{for all } r \in (0,L],$$

and

$$4\Big(F(L) + \psi_{1-\xi}(L,0)^{C}\mathcal{D}^{\zeta,\chi}F(L)\Big) < 1,$$

where $\Omega(r) = \{y(r) : y \in \Omega\}$, ${}^{C}\mathcal{D}^{\zeta,\chi}\Omega(r) = \{{}^{C}\mathcal{D}^{\zeta,\chi}y(r) : y \in \Omega\}$ and

$$F(r) = \frac{1}{\Gamma(\xi)\Gamma(\zeta)}\int_0^r\int_0^s \chi'(s)\chi'(\tau)\psi_{\zeta-1}(r,s)\psi_{\xi-1}(s,\tau)\,[\imath(\tau) + \jmath(\tau)]\,d\tau ds, \quad r \in (0,L].$$

$(\mathbf{H}_3)$

$$\left[\frac{A\psi_{\zeta+1}(L,0) + (\xi+\zeta+1)B\psi_{\zeta+2}(L,0) + f^*\psi_{\zeta+1}(L,0)}{\Gamma(\xi+\zeta+1)} + \frac{A\psi_1(L,0) + B\psi_{2-\xi}(L,0) + f^*\psi_1(r,0)}{\Gamma(\xi+1)}\right] < 1.$$

We define the operator $N : C^{\zeta}_{\xi,\chi}([0,L],\mathfrak{E}) \to C^{\zeta}_{\xi,\chi}([0,L],\mathfrak{E})$ by

$$\begin{aligned} Ny(r) &= \rho_2 + \rho_1 e^{-(\lambda+\mu)\psi(r)}\frac{\psi_{\zeta+\xi-1}(r,0)}{\Gamma(\zeta+\xi)} \\ &\quad + \frac{1}{\Gamma(\zeta)\Gamma(\xi)}\int_0^r\int_0^s \chi'(s)\chi'(\tau)\psi_{\zeta-1}(r,s)\psi_{\xi-1}(s,\tau)e^{-\lambda\psi_1(r,s)}e^{-\mu\psi_1(s,\tau)} \\ &\quad f(\tau, y(\tau), {}^{c}\mathcal{D}^{\zeta,\chi}y(\tau))d\tau ds, \end{aligned}$$

and the operator ${}^{CT}_{\mu}\mathcal{D}^{\zeta,\chi} : C_{\xi,\chi}([0,L],\mathfrak{E}) \to C_{\xi,\chi}([0,L],\mathfrak{E})$ by

$${}^{CT}_{\mu}\mathcal{D}^{\zeta,\chi}y(r) = \rho_1 e^{-\lambda\psi_1(r,0)}\psi_{\xi-1}(r,0) + \frac{1}{\Gamma(\xi)}\int_0^r \chi'(s)\psi_{\xi-1}(r,s)e^{-\lambda[\psi_1(r,s)]}y(s)ds,$$

The theorem below is the main result. Let

$$B = \left\{y \in C^{\zeta}_{\xi,\chi}([0,L],\mathfrak{E}) : \|y\|_{C^{\zeta}_{\xi,\chi}} \leq R\right\},$$

where R is a strictly positive real number.

Theorem 1.2 Suppose that the conditions $(\mathbf{H}_1) - (\mathbf{H}_3)$ are valid. If.

$(\mathbf{H}_4)$

$$\frac{\|\rho_2\| + \|\rho_1\|e^{-(\lambda+\mu)(\psi_1(L,0))}\frac{\psi_{\zeta+\xi-1}(L,0)}{\Gamma(\zeta+\xi)} + \frac{\|\rho_1\|}{\Gamma(\xi)}e^{-\lambda\psi_1(L,0)}}{\left[\frac{A\psi_{\zeta+1}(L,0)+(\xi+\zeta+1)B\psi_{\zeta+2}(L,0)+f^*\psi_{\zeta+1}(L,0)}{\Gamma(\xi+\zeta+1)} + \frac{A\psi_1(L,0)+B\psi_{2-\xi}(L,0)+f^*\psi_1(L,0)}{\Gamma(\xi+1)}\right]} \leq R$$

is valid. Then, the problem $(\mathfrak{P})$ has at least one solution.

Proof: Clearly, from lemma 2.8 fixed points of the operator N are solution of the problem $(\mathfrak{P})$. We shall use Meir-Keller fixed point theorem for condensing operator to prove that N has a fixed point. It will be proved in several steps.

First step. We begin to show that N is bounded. Let $y \in \mathcal{C}^{\zeta}_{\xi,\chi}([0,L],\mathfrak{E})$, from $(\mathbf{H}_2)$ and by an easy computation one deduces $Ny \in \mathcal{C}^{\zeta}_{\xi,\chi}([0,L],\mathfrak{E})$. Using $(\mathbf{H}_2)$ and Lemmas 2.3 and 2.4, for all $y \in B_{\varrho} = \left\{y \in \mathcal{C}^{\zeta}_{\xi,\chi}([0,L],\mathfrak{E}) : \|y\|_{\mathcal{C}^{\zeta}_{\xi,\chi}} < \varrho\right\}$ and $r \in (0,L]$ we get

$$
\begin{aligned}
Ny(r) &= \|\rho_2\| + \|\rho_1\| e^{-(\lambda+\mu)(\psi_1(r,0))} \frac{\psi_{\zeta+\xi-1}(r,0)}{\Gamma(\zeta+\xi)} \\
&+ \frac{1}{\Gamma(\zeta)\Gamma(\xi)} \int_0^r \int_0^s \chi'(s)\chi'(\tau)\psi_{\zeta-1}(r,s)\psi_{\xi-1}(s,\tau) e^{-\lambda\psi_1(r,s)} e^{-\lambda\psi_1(r,s)} \\
&(\|f(\tau,y(\tau),{}^c\mathcal{D}^{\zeta,\chi}y(\tau)) - f(\tau,0,0\|) + \|f,(\tau,0,0)\|)d\tau ds, \\
&\le \|\rho_2\| + \|\rho_1\| e^{-(\lambda+\mu)(\psi_1(r,0))} \frac{\psi_{\zeta+\xi-1}(r,0)}{\Gamma(\zeta+\xi)} \\
&+ \frac{1}{\Gamma(\zeta)\Gamma(\xi)} \int_0^r \int_0^s \chi'(s)\chi'(\tau)\psi_{\zeta-1}(r,s)\psi_{\xi-1}(s,\tau) e^{-\lambda\psi_1(r,s)} e^{-\lambda\psi_1(s,\tau)} \\
&(\|f(\tau,y(\tau),{}^c\mathcal{D}^{\zeta,\chi}y(\tau)) - f(\tau,0,0\|))d\tau ds \\
&+ \frac{1}{\Gamma(\zeta)\Gamma(\xi)} \int_0^r \int_0^s \chi'(s)\chi'(\tau)\psi_{\zeta-1}(r,s)\psi_{\xi-1}(s,\tau) e^{-\lambda\psi_1(r,s)} e^{-\lambda\psi_1(s,\tau)]} \|f(\tau,0,0)\| d\tau ds, \\
&\le \|\rho_2\| + \|\rho_1\| e^{-(\lambda+\mu)(\psi_1(r,0))} \frac{\psi_{\zeta+\xi-1}(r,0)}{\Gamma(\zeta+\xi)} + \frac{1}{\Gamma(\zeta)\Gamma(\xi)} \int_0^r \int_0^s \chi'(s)\chi'(\tau)\psi_{\zeta-1}(r,s)\psi_{\xi-1}(s,\tau) \\
&(A\|y\| + \psi_{1-\xi}(r,0)B\|{}^c\mathcal{D}^{\zeta,\chi}\|y\|)d\tau ds \\
&+ \frac{f^*}{\Gamma(\zeta)\Gamma(\xi)} \int_0^r \int_0^s \chi'(s)\chi'(\tau)\psi_{\zeta-1}(r,s)\psi_{\xi-1}(s,\tau)\|f(\tau,0,0)\| d\tau ds, \\
&\le \|\rho_2\| + \|\rho_1\| e^{-(\lambda+\mu)(\psi_1(r,0))} \frac{\psi_{\zeta+\xi-1}(r,0)}{\Gamma(\zeta+\xi)} \frac{A\rho\psi_{\xi+\zeta}(L,0)}{\Gamma(\xi+\zeta+1)} + \frac{B\rho\psi_{\zeta+1}(L,0)}{\Gamma(\zeta+2)} + \frac{f^* A\rho\psi_{\zeta+1}(L,0)}{\Gamma(\xi+\zeta+1)}
\end{aligned}
$$

So,

$$
\begin{aligned}
\|Ny\|_\infty \le \|\rho_2\| + \|\rho_1\| e^{-(\lambda+\mu)(\psi_1(L,0))} \frac{\psi_{\zeta+\xi-1}(L,0)}{\Gamma(\zeta+\xi)} + \frac{A\rho\psi_{\zeta+1}(L,0)}{\Gamma(\xi+\zeta+1)} \\
+ \frac{B\rho\psi_{\zeta+2}(L,0)}{\Gamma(\xi+\zeta+2)} + \frac{f^* A\rho\psi_{\zeta+1}(L,0)}{\Gamma(\xi+\zeta+1)} = M_1.
\end{aligned}
\tag{7}
$$

And we have also

$$
\begin{aligned}
&\|\psi_{1-\xi}(r,0)({}^c\mathcal{D}^{\zeta,\chi}N)y(r)\| \le \\
&\frac{\|\rho_1\|}{\Gamma(\xi)} e^{-\lambda\psi_1(r,0)} + \frac{\psi_{1-\xi}(r,0)}{\Gamma(\xi)} \int_0^r \chi'(s)\psi_{\xi-1}(r,s)e^{-\lambda\psi_1(r,s)} \|f(s,y(s),c\mathcal{D}^{\zeta,\chi}y(s))\| ds \\
&\le \frac{\|\rho_1\|}{\Gamma(\xi)} e^{-\lambda\psi_1(r,0)} + \frac{\psi_{1-\xi}(r,0)}{\Gamma(\xi)} \int_0^r \chi'(s)\psi_{\xi-1}(r,s)e^{-\lambda\psi_1(r,s)} \|f(s,y(s),c\mathcal{D}^{\zeta,\chi}y(s)) - f(s,0,0)\| ds \\
&+ \frac{\psi_{1-\xi}(r,0)}{\Gamma(\xi)} \int_0^r \chi'(s)\psi_{\xi-1}(r,s)e^{-\lambda\psi_1(r,s)} \|f(s,0,0)\| ds \\
&\le \frac{\|\rho_1\|}{\Gamma(\xi)} e^{-\lambda\psi_1(r,0)} + \frac{\psi_{1-\xi}(r,0)}{\Gamma(\xi)} \int_0^r \chi'(s)\psi_{\xi-1}(r,s)e^{-\lambda\psi_1(r,s)} (A\|y\| + \psi_{1-\xi}(r,0)B\|{}^c\mathcal{D}^{\zeta,\chi}\|y\|) ds \\
&+ \frac{\psi_{1-\xi}(r,0)}{\Gamma(\xi)} \int_0^r \chi'(s)\psi_{\xi-1}(r,s)e^{-\lambda\psi_1(r,s)} \|f(s,0,0)\| ds \\
&\le \frac{\|\rho_1\|}{\Gamma(\xi)} e^{-\lambda\psi_1(r,0)} + \frac{A\rho\psi_1(r,0)}{\Gamma(\xi)} + \frac{B\rho\psi_{2-\xi}(r,0)}{\Gamma(\xi)} + \frac{f^*\rho\psi_1(r,0)}{\Gamma(\xi)}
\end{aligned}
$$

So,

$$\|{}^c\mathcal{D}^{\zeta,\chi}N)y\|_{C_{\xi,\chi}} \leq \frac{\|\rho_1\|}{\Gamma(\xi)}e^{-\lambda\psi_1(L,0)} + \frac{A\rho\psi_1(L,0)}{\Gamma(\xi)} + \frac{B\rho\psi_{2-\xi}(L,0)}{\Gamma(\xi)} + \frac{f^*\rho\psi_1(L,0)}{\Gamma(\xi)} = M_2, \tag{8}$$

From (7) and (8), we get

$$\|(Ny)\|_{C^{\zeta}_{\xi,\chi}} \leq M = M_1 + M_2.$$

Second step. We prove that N is continuous. Let $\{y_n\}_{n\in N}$ converges to y in $C^{\zeta}_{\xi,\chi}([0,L],\mathfrak{E})$, from the fact that f is a Carathéodoty function and using condition (H_1) we deduce that Ny_n converge to Ny in $C([0,L],\mathfrak{E})$ Similarly $\psi_{1-\xi}(r,0)\|({}^c\mathcal{D}^{\zeta,\chi}N)y_n$ converge to $\psi_{1-\xi}(r,0)\|({}^c\mathcal{D}^{\zeta,\chi}N)y$ in $C_{\xi,\chi}$ Finally, Ny_n converge to Ny in $C^{\zeta}_{\xi,\chi}([0,L],\mathfrak{E})$.

Third step. We aim to show that NB is equi-continuous for all bounded subset B of $C^{\zeta}_{\xi,\chi}([0,L],\mathfrak{E})$, let B_ρ be the subset which was previously defined. It suffices to prove that NB_ρ and $({}^c\mathcal{D}^{\zeta,\chi}N)B_\rho$ are equicontinuous respectively in $C([0,L],\mathfrak{E})$ and in $C_{\xi,\chi}([0,L],\mathfrak{E})$. Let $y\in B_\rho$ and $r_1,r_2\in(0,L]$ with $r_1<r_2$. First of all, we have

$$\begin{aligned}
\|Ny(r_2)-Ny(r_1)\| &\leq \frac{\|\rho_1\|e^{-(\lambda+\mu)\|\psi_1(r_1,0)-\psi_1(r_2,0)\|}\left(\psi_{\xi+\zeta-1}(r_2,0)-\psi_{\xi+\zeta-1}(r_1,0)\right)}{\Gamma(\xi+\zeta)}\\
&+\frac{1}{\Gamma(\xi)\Gamma(\zeta)}\int_0^{r_1}\int_0^s \chi'(s)\chi'(\tau)\left[\psi_{\zeta-1}(r_1,s)-\psi_{\zeta-1}(r_2,s)\right]\psi_{\xi-1}(s,\tau)\\
&|e^{-\lambda\psi_1(r_1,s)}-e^{-\lambda\psi_1(r_2,s)}|e^{-\lambda\psi_1(s,\tau)}\|f\left(\tau,y(\tau),{}^c\mathcal{D}^{\zeta,\chi}y(\tau)\right)\|d\tau ds\\
&+\frac{1}{\Gamma(\xi)\Gamma(\zeta)}\int_{r_1}^{r_2}\int_0^s \chi'(s)\chi'(\tau)\psi_{\zeta-1}(r_2,s)\psi_{\xi-1}(s,\tau)\\
&e^{-\lambda\psi_1(r_2,s)}|e^{-\lambda\psi_1(s,\tau)}\|f\left(\tau,y(\tau),{}^c\mathcal{D}^{\zeta,\chi}y(\tau)\right)\|d\tau ds\\
&\leq \frac{\|\rho_1\|\left(\psi_{\xi+\zeta-1}(r_2,0)-\psi_{\xi+\zeta-1}(r_1,0)\right)\left[e^{-(\lambda+\mu)\|\psi_1(r_1,0)-\psi_1(r_2,0)}\right]\|}{\Gamma(\xi+\zeta)}\\
&+\frac{1}{\Gamma(\zeta)\Gamma(\xi)}\int_0^{r_1}\int_0^s \chi'(s)\chi'(\tau)\left[\psi_{\zeta-1}(r_1,s)-\psi_{\zeta-1}(r_2,s)\right]\psi_{\xi-1}(s,\tau)\\
&(A\|y\|+|\psi_{1-\xi}(r_1,0)-\psi_{1-\xi}(r_2,0|b(r)\|{}^c\mathcal{D}^{\zeta,\chi}\|y\|)d\tau ds\\
&+\frac{f^*}{\Gamma(\zeta)\Gamma(\xi)}\int_0^{r_1}\int_0^s \chi'(s)\chi'(\tau)|\psi_{\zeta-1}(r_1,s)-\psi_{\zeta-1}(r_2,s)|\psi_{\xi-1}(s,\tau)d\tau ds\\
&+\frac{1}{\Gamma(\zeta)\Gamma(\xi)}\int_{r_1}^{r_2}\int_0^s \chi'(s)\chi'(\tau)|\psi_{\zeta-1}(r_2,s)|\psi_{\xi-1}(s,\tau)\\
&(a(r)\|y\|+\psi_{1-\xi}(r_2,0)B\|{}^c\mathcal{D}^{\zeta,\chi}\|y\|)d\tau ds\\
&+\frac{f^*}{\Gamma(\zeta)\Gamma(\xi)}\int_{r_1}^{r_2}\int_0^s \chi'(s)\chi'(\tau)\psi_{\zeta-1}(r_2,s)\psi_{\xi-1}(s,\tau)d\tau ds\\
&\leq \frac{\|\rho_1\|\left(\psi_{\xi+\zeta-1}(r_2,0)-\psi_{\xi+\zeta-1}(r_1,0)\right)\left[e^{-(\lambda+\mu)\|\psi_1(r_1,0)-\psi_1(r_2,0)}\right]\|}{\Gamma(\xi+\zeta)}\\
&+\frac{|\psi_{\zeta+\xi}(r_1,0)-\psi_{\zeta+\xi}(r_1,0)|}{\Gamma(\xi+\zeta+1)}(a(r)\rho+|\psi_{1-\xi}(r_1,0)-\psi_{1-\xi}(r_2,0)|B\rho\\
&+f^*\frac{|\psi_{\zeta+\xi}(r_1,0)-\psi_{\zeta+\xi}(r_1,0)|}{\Gamma(\xi+\zeta+1)}+(A\rho+\psi_{1-\xi}(r_2,0)B\|\rho)\\
&+\frac{1}{\Gamma(\zeta)\Gamma(\xi)}\int_{r_1}^{r_2}\int_0^s \chi'(s)\chi'(\tau)|\psi_{\zeta-1}(r_2,s)|\psi_{\xi-1}(s,\tau)\\
&+\frac{f^*}{\Gamma(\zeta)\Gamma(\xi)}\int_{r_1}^{r_2}\int_0^s \chi'(s)\chi'(\tau)\psi_{\zeta-1}(r_2,s)\psi_{\xi-1}(s,\tau)d\tau ds
\end{aligned}$$

Taking r_2 tends toward r_1, we get that, the last formula tends to zero. Then NB_ρ is equicontinuous in $\mathcal{C}([0, L], \mathfrak{E})$.

And, we also have

$$
\begin{aligned}
&\|\psi_{1-\xi}(r_2, 0)\left({}^c\mathcal{D}^{\zeta,\chi}N\right)y(r_2) - \psi_{1-\xi}(r_1, 0)\left({}^c\mathcal{D}^{\zeta,\chi}N\right)(y)(r_1)\| \\
&\le \|\rho_1\| e^{-\lambda\psi_1(r_1,0)} - e^{-\lambda\psi_1(r_2,0)}| + \\
&\frac{\psi_{1-\xi}(r_1, 0)}{\Gamma(\xi)}\int_0^{r_1} \chi'(s)\left[\psi_{\xi\ 1}(r_1, s) - \psi_{\zeta-1}(r_2, s)\right]\|f\left(s, y(s), {}^c\mathcal{D}^{\zeta,\chi}y(s)\right)\|ds \\
&+\frac{\left[\psi_{1-\xi}(r_2, 0) - \psi_{1-\xi}(r_1, 0)\right]}{\Gamma(\xi)}\int_0^{r_1} \chi'(s)\psi_{\xi-1}(r_2, s)\|f\left(s, y(s), {}^c\mathcal{D}^{\zeta,\chi}y(s)\right)\|ds \\
&+\frac{\psi_{1-\xi}(r_2, 0)}{\Gamma(\xi)}\int_{r_1}^{r_2} \chi'(s)\psi_{\xi-1}(r_2, s)\|f\left(s, y(s), {}^c\mathcal{D}^{\zeta,\chi}y(s)\right)\|ds \\
&\le \|\rho_1\| e^{-\lambda\psi_1(r_1,0)} - e^{-\lambda\psi_1(r_2,0)}| + \\
&\left(A\rho + B\rho|\psi_{1-\xi}(r_1, 0) - \psi_{1-\xi}(r_2, 0)|\right)\frac{\psi_{1-\xi}(r_1, 0)}{\Gamma(\xi)}\int_0^{r_1} \chi'(s)\left[\psi_{\xi-1}(r_1, s) - \psi_{\xi-1}(r_2, s)\right]ds \\
&f^*\frac{\psi_{1-\xi}(r_1, 0)}{\Gamma(\xi)}\int_0^{r_1} \chi'(s)\left[\psi_{\xi-1}(r_1, s) - \psi_{\xi-1}(r_2, s)\right]ds \\
&+\left(A\rho + B\rho|\psi_{1-\xi}(r_1, 0) - \psi_{1-\xi}(r_2, 0)|\right)\frac{\left[\psi_{1-\xi}(r_2, 0) - \psi_{1-\xi}(r_1, 0)\right]}{\Gamma(\xi)}\int_0^{r_1} \chi'(s)\psi_{\xi-1}(r_2, s)ds \\
&+f^*\frac{\left[\psi_{1-\xi}(r_2, 0) - \psi_{1-\xi}(r_1, 0)\right]}{\Gamma(\xi)}\int_0^{r_1} \chi'(s)\psi_{\xi-1}(r_2, s)ds \\
&+\left(A\rho + B|\psi_{1-\xi}(r_1, 0) - \psi_{1-\xi}(r_2, 0)|\right)\frac{\psi_{1-\xi}(r_2, 0)}{\Gamma(\xi)}\int_{r_1}^{r_2} \chi'(s)\psi_{\xi-1}(r_2, s)ds \\
&+f^*\frac{\psi_{1-\xi}(r_2, 0)}{\Gamma(\xi)}\int_{r_1}^{r_2} \chi'(s)\psi_{\xi-1}(r_2, s)ds \\
&\le \|\rho_1\| e^{-\lambda\psi_1(r_1,0)} - e^{-\lambda\psi_1(r_2,0)}| + \\
&\left(A\rho + B\rho|\psi_{1-\xi}(r_1, 0) - \psi_{1-\xi}(r_2, 0)|\right)\frac{\psi_{1-\xi}(r_1, 0)}{\Gamma(\xi)}\left[\psi_{\xi-1}(r_2, 0) - \psi_{\xi-1}(r_1, 0) - \psi_{\xi-1}(r_2, r_1)\right] \\
&f^*\frac{\psi_{1-\xi}(r_1, 0)}{\Gamma(\xi)}\left[\psi_{\xi-1}(r_2, 0) - \psi_{\xi-1}(r_1, 0) - \psi_{\xi-1}(r_2, r_1)\right] \\
&+\left(A\rho + B\rho|\psi_{1-\xi}(r_1, 0) - \psi_{1-\xi}(r_2, 0)|\right)\frac{\left[\psi_{1-\xi}(r_2, 0) - \psi_{1-\xi}(r_1, 0)\right]}{\Gamma(\xi)}\left[\psi_{\xi-1}(r_2, r_1 - \psi_{\xi-1}(r_2, 0)\right] \\
&+f^*\frac{\left[\psi_{1-\xi}(r_2, 0) - \psi_{1-\xi}(r_1, 0)\right]}{\Gamma(\xi)}\left[\psi_{\xi-1}(r_2, r_1 - \psi_{\xi-1}(r_2, 0)\right]ds \\
&+\left(A\rho + B|\psi_{1-\xi}(r_1, 0) - \psi_{1-\xi}(r_2, 0)|\right)\frac{\psi_{1-\xi}(r_2, 0)}{\Gamma(\xi)}\left[\psi_{\xi-1}(r_2, r_1)\right] \\
&+f^*\frac{\psi_{1-\xi}(r_2, 0)}{\Gamma(\xi)}\left[\psi_{\xi-1}(r_2, r_1)\right]
\end{aligned}
$$

Taking r_2 tends toward r_1, we get that, the last inequality tends to zero. Then, $\left({}^c\mathcal{D}^{\zeta,\chi}N\right)B_\rho$ is equicontinuous in $\mathcal{C}_{\xi,\chi}([0, L], \mathfrak{E})$.

Final step. We shall prove that N verifies the assumptions of theorem 1.1.

First, we demonstrate that N is defined from B to B, Indeed, for any $y \in B$, by above condition $(\mathbf{H}_2)$ and by an easy computation one can deduce

$$
\begin{aligned}
&\|Ny(r)\| + \|\psi_{1-\xi}(r,0)\left({}^{c}\mathcal{D}^{\zeta,\chi}N\right)y(r)\| \\
&\leq \|\rho_2\| + \|\rho_1\| e^{-(\lambda+\mu)(\psi_1(L,0))}\frac{\psi_{\zeta+\xi-1}(L,0)}{\Gamma(\zeta+\xi)} + \frac{\|a\|}{\Gamma(\xi)}e^{-\lambda\psi_1(L,0)} \\
&\left[\frac{A\psi_{\zeta+1}(L,0) + (\xi+\zeta+1)B\psi_{\zeta+2}(L,0) + f^*\psi_{\zeta+1}(L,0)}{\Gamma(\xi+\zeta+1)}\right. \\
&\left.+\frac{A\psi_1(L,0) + B\psi_{2-\xi}(r,0) + f^*\psi_1(L,0)}{\Gamma(\xi+1)}\right]R
\end{aligned}
$$

From $(\mathbf{H}_3)$ and $(\mathbf{H}_4)$, we obtain

$$
\forall y \in B : \|Ny\|_{C^{\zeta}_{\xi,\chi}} < R.
$$

Consequently, N is defined from B to B. Taking $D = \overline{conv}(NB)$, where $\overline{conv}(NB)$ is the closure of the convex hull of NB. Since $ND \subset NB \subset D$, then D is a subset closed, bounded and convex of B and N remains defined from D to D.

We point out $\vartheta_{(\xi,\chi)}$ the Kuratowski measure of non-compactness defined on any bounded subset of $C^{\zeta}_{\xi,\chi}([0,L],\mathfrak{E})$. one has facilely the following inequality

$$
\vartheta_{(\xi,\chi)}(D) \leq \sup_{r\in[0,L]} \vartheta(D(r)) + \sup_{r\in(0,L]} \vartheta\left(\psi_{1-\xi}(r,0)\,{}^{C}\mathcal{D}^{\zeta,\chi}D(r)\right) \leq 2\vartheta_{(\xi,\chi)}(D), \tag{9}
$$

$D(r) = \{y(r) : y \in D\}$ and ${}^{C}\mathcal{D}^{\zeta,\chi}D(r) = \left\{{}^{c}\mathcal{D}^{\zeta,\chi}y(r) : y \in D\right\}$.

Next, it remains to prove that N is a Meir-Keeler condensing operator with the measure of non-compactness $\vartheta_{(\xi,\chi)}$, this is equivalent to prove the implication below

$$
\forall \epsilon > 0, \exists \varrho(\epsilon) > 0 : \epsilon \leq \vartheta_{(\xi,\chi)}(V) < \epsilon + \varrho \Rightarrow \vartheta_{(\xi,\chi)}(NV) < \epsilon, \quad \text{for any} \quad V \subset D. \tag{10}
$$

Let $\epsilon > 0$, $V \subset D$. From Lemmas 2.5, 2.6, 2.7, $(\mathbf{H}_2)$, the inequality (9) and the steps above, it follows then that there exists a sequence $\{\mu_n\}_{n=0}^{\infty} \subset V$ such that

$$
\begin{aligned}
&\vartheta(NV(r)) + \vartheta\left(\psi_{1-\xi}(r,0)\left({}^{C}\mathcal{D}^{\zeta,\chi}N\right)V(r)\right) \leq \frac{\epsilon}{2} \\
&+\frac{2}{\Gamma(\zeta)\Gamma(\xi)}\vartheta\left\{\int_0^r\int_0^s \chi'(s)\chi'(\tau)\psi_{\zeta-1}(r,s)\psi_{\xi-1}(s,\tau)f(\tau,\mu_n(\tau),{}^{C}\mathcal{D}^{\zeta,\chi}\mu_n(\tau))d\tau ds,\ n\in\mathbb{N}\right\} \\
&+\frac{2\psi_{1-\xi}(r,0)}{\Gamma(\xi)}\vartheta\left\{\int_0^r \chi'(s)\psi_{\xi-1}(r,s)f(s,\mu_n(s),{}^{C}\mathcal{D}^{\zeta,\chi}\mu_n(s))ds,\ n\in\mathbb{N}\right\} \\
&\leq \frac{\epsilon}{2} + \frac{2}{\Gamma(\zeta)\Gamma(\xi)}\int_0^r\int_0^s \chi'(s)\chi'(\tau)\psi_{\zeta-1}(r,s)\psi_{\xi-1}(s,\tau)\vartheta\left\{f(\tau,\mu_n(\tau),{}^{C}\mathcal{D}^{\zeta,\chi}\mu_n(\tau)),\ n\in\mathbb{N}\right\}d\tau ds \\
&+\frac{2\psi_{1-\xi}(r,0)}{\Gamma(\xi)}\int_0^r \chi'(s)\psi_{\xi-1}(r,s)\vartheta\left\{f(s,\mu_n(s),{}^{C}\mathcal{D}^{\zeta,\chi}\mu_n(s)),\ n\in\mathbb{N}\right\}ds \\
&\leq \frac{\epsilon}{2} + 4\vartheta_{(\xi,\chi)}(V)\left\{F(L) + \psi_{1-\xi}(L,0)\,{}^{C}\mathcal{D}^{\zeta,\chi}F(L)\right\}.
\end{aligned}
$$

From (9), we know that

$$
\vartheta_{(\xi,\chi)}(NV) \leq \frac{\epsilon}{2} + 4\vartheta_{(\xi,\chi)}(V)\left\{F(L) + \psi_{1-\xi}(L,0)\,{}^{C}\mathcal{D}^{\zeta,\chi}F(L)\right\}.
$$

If

$$\vartheta_{(\xi,\chi)}(NV) \leq \frac{\epsilon}{2} + 4\vartheta_{(\xi,\chi)}(V)\{F(L) + \psi_{1-\xi}(L,0)^{C}\mathcal{D}^{\zeta,\chi}F(L)\} < \epsilon,$$

this implies that

$$\vartheta_{(\xi,\chi)}(V) < \frac{1}{8\left(F(L) + \psi_{1-\xi}(L,0)^{c}\mathcal{D}^{\zeta,\chi}F(L)\right)}\epsilon,$$

so that implication (10) is fulfilled, we take

$$\varrho = \frac{1 - 8\left(F(L) + \psi_{1-\xi}(L,0)^{c}\mathcal{D}^{\zeta,\chi}F(L)\right)}{8\left(F(L) + \psi_{1-\xi}(L,0)^{C}\mathcal{D}^{\zeta,\chi}F(L)\right)}\epsilon.$$

Consequently, N is a Meir-Keeler condensing operator with $\vartheta_{(\xi,\chi)}$, finally all the assumptions of the theorem 1.1 are satisfied. Then, the problem $(\mathfrak{P})$ has at least one solution.

4. Example

We shall now give now an example as application to our main result.

$$^{rl}\mathcal{D}^{\frac{1}{2},\frac{1}{2},\chi}\left({}^{c}\mathcal{D}^{\frac{1}{4},\frac{1}{2},\chi}\right)y(r) = \frac{1}{40(1+r^2)}\left(y_n(r) + \frac{\sqrt{\arctan r}}{e^{5\sqrt{2}\arctan r}}\left({}^{c}\mathcal{D}^{\frac{3}{4},\frac{1}{2},\chi}\right)y_n(r)\right)_{n=1}^{\infty}, \quad r \in (0,1], \tag{11}$$

$$\mathfrak{I}^{\frac{1}{2}}_{0^+}\left({}^{c}\mathcal{D}^{\frac{1}{4},\frac{1}{2},\chi}\right)y(0^+) = (1, 0, \dots, 0, \dots) \tag{12}$$

$$y(0) = (1, 0, \dots, 0, \dots). \tag{13}$$

where $\chi(r) = \frac{4}{\pi}\arctan r$. Let

$$E = \left\{(y_1, y_2, \dots, y_n, \dots) : \sup_n |y_n| < \infty\right\},$$

with the norm $\|y\| = \sup_n |y_n|$, then $(E, \|.\|)$ consists a Banach space, by comparing with the **(P)**, we notice that

$$\xi = \lambda = \frac{1}{2}, \zeta = \frac{1}{4} \text{ and } f\left(r, y(r), {}^{c}\mathcal{D}^{\frac{1}{4},\frac{1}{2},\chi}y(r)\right) = \left(f\left(r, y_1(r), {}^{c}\mathcal{D}^{\frac{1}{4},\frac{1}{2},\chi}y_1(r)\right), \dots, f\left(r, y_n(r), {}^{c}\mathcal{D}^{\frac{1}{4},\frac{1}{2},\chi}y_n(r)\right), \dots\right),$$

where

$$f\left(r, y_n(r), {}^{c}\mathcal{D}^{\frac{1}{4},\frac{1}{2},\chi}y_n(r)\right) = \frac{1}{40(1+r^2)}y_n(r) + \frac{\sqrt{\arctan r}}{40(1+r^2)e^{5\sqrt{2}\arctan r}}\,{}^{c}\mathcal{D}^{\frac{1}{4},\frac{1}{2},\chi}y_n(r), n \in N^*.$$

Clear that $f : (0,1] \times E \times E \to E$ is continuous and

$$\|f\left(r,y(r),{}^{c}\mathcal{D}^{\frac{1}{4},\frac{1}{2};\chi}y(r)\right)\| \leq \frac{1}{40(1+r^2)}\|y(r)\| + \frac{\sqrt{\arctan r}}{40(1+r^2)e^{5\sqrt{2}\arctan r}}\|{}^{c}\mathcal{D}^{\frac{1}{4},\frac{1}{2};\chi}y(r)\|.$$

Hence, $(\mathbf{H_1})$ and $(\mathbf{H_2})$ are satisfied. Next, for any bounded set $B \subset C^{\zeta}_{\xi,\chi}([0,1],E)$, we have

$$f\left(r,B(r),{}^{c}\mathcal{D}^{\frac{1}{4},\frac{1}{2};\chi}B(r)\right) = \frac{1}{40(1+r^2)}B(r) + \frac{\sqrt{\arctan r}}{40(1+r^2)e^{5\sqrt{2}\arctan r}}\,{}^{c}\mathcal{D}^{\frac{1}{4},\frac{1}{2};\chi}B(r).$$

Then

$$\vartheta(f\left(r,B(r),{}^{c}\mathcal{D}^{\frac{1}{4},\frac{1}{2};\chi}B(r)\right) \leq \frac{1}{40(1+r^2)}\vartheta(B(r)) + \frac{1}{40(1+r^2)}\vartheta\left(\sqrt{\arctan r}\times{}^{c}\mathcal{D}^{\frac{1}{4},\frac{1}{2};\chi}B(r)\right),$$

since $4\left(F(1)+\chi_{1-\xi}(10)^{c}\mathcal{D}^{\frac{1}{4},\frac{1}{2};\chi}F(1)\right) \leq <1$. So, $(\mathbf{H_3})$ holds. Finally, we check $(\mathbf{H_4})$, we have it can be written as

$$\left[\frac{\frac{1}{40}\psi_{\frac{5}{4}}(1,0)+\left(\frac{7}{4}\right)\frac{1}{40}\psi_{\frac{9}{4}}(1,0)+}{\Gamma\left(\frac{7}{4}\right)} + \frac{\frac{1}{40}\psi_{1}(1,0)+\frac{1}{40}\psi_{\frac{3}{2}}(1,0)}{\Gamma\left(\frac{3}{2}\right)}\right] < 1.$$

An easy computation yields that that condition $(\mathbf{H_4})$ is satisfied.

Therefore, Theorem 2 ensures that problem (4)-(11) has a solution.

Acknowledgements

The authors would like to express their thanks to the editor and anonymous referees for his/her suggestions and comments that improved the quality of the paper.

The authors are grateful Prof Nina Miocevic Publishing Process Managerto, The Editor and anonymous referees and the editor for carefully reading this manuscript and giving valuable suggestions improving the quality of the paper.

Funding

The authors were supported in part by Project of research COOL03UN140120220002 supported by high ministry of education.

Conflict of interest

The authors declare no conflict of interest.

Use of AI tools declaration

The authors declare they have not used Artificial Intelligence (AI) tools in the creation of this article.

Data availability statements

The study protocol (hypotheses, methods, and analysis plan) was preregistered on June 2024. The lemmas and theorems and analysis tools needed for this manuscript are explicitly acknowledged.

Author details

Amira M'boudi and Benaouda Hedia*
Department of Mathematics, Faculty of Mathematics and Computers Sciences, University Ibn Khaldoun Tiaret, Algeria

*Address all correspondence to: b_hedia@univ-tiaret.dz

IntechOpen

© 2025 The Author(s). Licensee IntechOpen. This chapter is distributed under the terms of the Creative Commons Attribution License (http://creativecommons.org/licenses/by/4.0), which permits unrestricted use, distribution, and reproduction in any medium, provided the original work is properly cited.

References

[1] Beddani M, Hedia B. Solution sets for fractional differential inclusions. Journal of Fractional Calculus & Applications. 2019;**10**(2):273-289

[2] Beddani M, Hedia B. Existence result for a fractional differential equation involving special derivative. Moroccan Journal of Pure and Applied Analysis. 2022;**8**(1):67-77

[3] Beddani M, Hedia B. Existence result for fractional differential equation on unbounded domain. Kragujevac Journal of Mathematics. 2024;**48**(5):755-766

[4] Berrabah FZ, Boukrouche M, Hedia B. Fractional derivatives with respect to time for non-classical heat problem. Journal of Mathematical Physics, Analysis, Geometry. 2021;**17**:30-53

[5] Hedia B. Non-local Conditions for Semi-Linear Fractional Differential Equations with Hilfer Derivative, Springer Proceeding in Mathematics and Statistics 303, ICFDA 2018, Amman, Jordan, July 16-18. pp. 69-83

[6] Hedia B. Structure of solution sets for fractional partial integro-differential equations. In: Differential and Difference Equations with Applications, ICDDEA 2019, Lisbon, Portugal, July 1-5. 2020. pp. 311-322. DOI: 10.1007/978-3-030-56323-3 24

[7] Hedia B. Topological structure of solution sets for hybrid impulsive fractional differential inclusion in Banach algebra. Journal of Computational Mathematics. 2022;**66**(1):348-371. DOI: 10.26524/cm138

[8] Kilbas AA, Srivastava HM, Trujillo JJ. Theory and Applications of Fractional Differential Equations. Amsterdam: North-Holland Mathematics Studies; 2006

[9] Miller KS, Ross B. An Introduction to the Fractional Calculus and Differential Equations. New York: John Wiley; 1993

[10] Podlubny I. Fractional Differential Equations. San Diego: Academic Press; 1999

[11] Samko SG, Kilbas AA, Marichev OI. Fractional Integrals and Derivatives Theory and Applications. Yverdon: Gordon and Breach; 1993

[12] Zhou Y. Basic Theory of Fractional Differential Equations. Singapore: World Scientific; 2014. pp. 72-91

[13] Berrabah FZ, Hedia B, Kirane M. Lyapunov- and Hartman U Wintner-type inequalities for a nonlinear fractional BVP with generalized-Hilfer derivative. Mathematical Methods in Applied Sciences. 2021;**44**:2637-2649

[14] Ledesma Torres C, Nyamoradi NEMAT, Pichardo Diestra O. Boundary value problem with tempered fractional derivatives. 13 Jul 2023. DOI: 10.13140/RG.2.2.13716.88967 [Preprint]

[15] Li C, Deng W, Zhao L. Well-posedness and numerical algorithm for the tempered fractional differential equations. Discrete and Continuous Dynamical Systems - Series B. 2019;**24**:1989-2015

[16] Pandey P, Pandey R, Yadav S, Agrawal O. Variational approach for tempered fractional Sturm-Liouville problem. International Journal of Applied and Computational Mathematics. 2021;7:51

[17] Torres Ledesma C, Cuti Gutierrez H, Cuti Gutierrez HA, Avalos Rodrìguez JP, Zubiaga Vera W. Some boundedness

results for Riemann-Liouville tempered fractional integrals. Fractional Calculus and Applied Analysis. 2024;**27**(1)

[18] Torres Ledesma C, Cesar E, et al. Boundary value problem with tempered fractional derivatives and oscillating term. Journal of Pseudo-Differential Operators and Applications. 2023;**14**(4):62

[19] Sousa JVdC, de Oliveira EC. On the ψ-Hilfer fractional derivative. Communications in Nonlinear Science and Numerical Simulation. 2018; **60**:72-91

[20] Berrabah FZ, Hedia B, Henderson J. Fully Hadamard and Erdélyi-Kober-type integral boundary value problem of a coupled system of implicit differential equations. Turkish Journal of Mathematics. 2019;**43**:1308-1329

[21] Benia K, Ait Dads E, Beddani M, Hedia B. Nonlocal fractional differential equation on the half line in Banach space. Advances in the Theory of Nonlinear Analysis and Its Application. 2022;**6**(1):118-134

[22] Benia K, Beddani M, Feckan M, Hedia B. Existence result for a problem involving ψ-Riemann Liouville fractional derivative on unbbounded domain. Journal of Differential Equations and Application. 2022;**14**(1):83-97

[23] Hedia B, Benia K. Hybrid implicit fractional integro-differential equations with Hadamard integral boundary conditions. Dynamics of Continuous, Discrete and Impulsive Systems. 2021; **28**:207-222

Chapter 3

Existence and Ulam Stability for Fractional q-Difference Equations with Integral Boundary Conditions

Nadia Allouch and Samira Hamani

Abstract

This work discusses the existence and uniqueness of solutions to the boundary value problem for fractional q-difference equations, which involves the Caputo's fractional q-derivative with nonlinear integral conditions, by using some standard fixed point theorems. In addition, we study the Ulam-Hyers and Ulam-Hyers-Rassias stabilities of its solutions. Finally, we present two examples to explain our results.

Keywords: fractional q-difference equations, Caputo fractional q-derivative, fixed point, integral boundary conditions, Ulam stability

1. Introduction

Recently, many researchers have addressed differential equations of fractional order in various fields of science and engineering, such as physics, chemistry, biology, and economics. For more details, see the books of Samko et al. [1], Miller et al. [2], Podlubny [3], Hilfer [4], Kilbas et al. [5], and Tarasov [6]. Boundary value problems with integral boundary conditions constitute an important class of problems. Moreover, considerable attention has been given to the existence of solutions to the boundary value problems for fractional differential equations by a number of authors, see the papers of Agrawal et al. [7], Benchohra et al. [8], and Afshari et al. [9].

q-Difference calculus or quantum calculus is an old subject that was initially developed in the twentieth century by Jackson [10, 11]; it has become a link between physics and mathematics. The fundamental concepts of q-difference calculus can be found in Refs. [12, 13]. Furthermore, the fractional q-difference calculus was initially proposed by Al-Salam [14] and Agarwal [15] and has attracted the interest of numerous academics due to its use in mathematical modeling across a variety of fields, including the technical sciences, engineering, physics, and biomathematics.

Fractional q-difference equations that include Caputo's fractional q-derivative also play a crucial role in describing many phenomena in various areas. Consequently, the boundary value problems for fractional q-difference equations have gained importance by many authors where they investigated the existence of its solutions; for more information, see the works of Abbas et al. [16–18], Ahmed et al. [19, 20], Allouch et al. [21–23], and Annaby et al. [24].

 IntechOpen

The stability problem of differential equations was formulated and discussed by Ulam and Hyers; see the papers of Hyers [25], Ulam [26]. In 1978, Rassias [27] provided a generalization of the Hyers theorem, which allows the Cauchy difference to be unbounded. Following this result, many mathematicians were attracted and motivated to investigate the Ulam-Hyers stability and Ulam-Hyers-Rassias stability of fractional differential equations. See the works of Abbas et al. [17, 18, 28], Jung [29] and Rassias [27], Rus [30, 31], Taieb et al. [32], and Houas et al. [33].

Benchohra et al. [8] studied the existence and uniqueness of solutions to the boundary value problem for fractional order differential equations with nonlinear integral conditions of the following form:

$$\begin{cases} {}^{C}D^{\alpha}y(t) = f(t,y(t)), t\in J = [0,T], \quad 1<\alpha\leq 2, \\ y(0) - y'(0) = \displaystyle\int_0^T g(s,y(s))ds, \\ y(T) + y'(T) = \displaystyle\int_0^T h(s,y(s))ds, \end{cases} \tag{1}$$

where ${}^{C}D^{\alpha}$ is the Caputo's fractional derivative, and f, g and $h : J \times \mathbb{R} \to \mathbb{R}$ are given continuous functions.

In light of the previously described research, we address in this study the existence, uniqueness, and stability of solutions to the boundary value problem for fractional q-difference equations with nonlinear integral conditions of the following type:

$$\begin{cases} \left({}^{C}D_q^{\alpha}y\right)(t) = \phi(t,y(t)), t\in J = [0,T], \quad 1<\alpha\leq 2, \\ y(0) - y'(0) = \displaystyle\int_0^T \varphi(s,y(s))ds, \\ y(T) + y'(T) = \displaystyle\int_0^T \psi(s,y(s))ds, \end{cases} \tag{2}$$

where $q\in(0,1)$, $T>0$, ${}^{C}D_q^{\alpha}$ is the Caputo's fractional q-derivative of order α, and $\phi : J \times \mathbb{R} \to \mathbb{R}$ is a given function and $\varphi, \psi : J \times \mathbb{R} \to \mathbb{R}$ are continuous functions.

The format of the work is as follows: In Section 2, we review notations, basic definitions, and some properties of the fractional q-calculus. After that in Section 3, we present three main results about the existence and uniqueness of solutions of the boundary value problem (2), which is based on the fixed point theorems (Banach, Schaefer's, and Leray-Schauder nonlinear alternative). In addition, we define and study the Ulam-Hyers and Ulam-Hyers-Rassias stabilities of the boundary value problem (2). Lastly, we provide two examples in Section 4 to highlight the results obtained.

2. Preliminaries

This section provides basic notations, definitions, and facts that will be used in the rest of this work.

Let $T>0$ and define $J:=[0,T]$. Consider $C(J,\mathbb{R})$ the Banach space of continuous functions $y : J \to \mathbb{R}$, equipped with the norm

$$\|y\|_{\infty} = \sup_{t\in J}|y(t)|. \tag{3}$$

Let $L^1(J, \mathbb{R})$, the Banach space of measurable functions $y : J \to \mathbb{R}$, which are Lebesgue integrable, with the norm

$$\|y\|_{L^1} = \int_J |y(t)|dt. \tag{4}$$

Now, we review definitions and some properties of fractional q-calculus [12, 13]. For $q \in (0,1)$ and let $a \in \mathbb{R}$, we set

$$[a]_q = \frac{1-q^a}{1-q}. \tag{5}$$

For all $a, b \in \mathbb{R}$ and $n \in \mathbb{N}$, the q-analogue of the power $(a-b)^{(n)}$ is expressed by

$$(a-b)^{(n)} = \prod_{k=0}^{n-1}\left(a - bq^k\right). \tag{6}$$

In general,

$$(a-b)^{(\alpha)} = a^\alpha \prod_{k=0}^{\infty}\left(\frac{a-bq^k}{a-bq^{k+\alpha}}\right), \quad a, b, \alpha \in \mathbb{R}. \tag{7}$$

Note that if $b = 0$, then $a^{(\alpha)} = a^\alpha$.

Definition 1.1 [13]: The q-gamma function is defined by

$$\Gamma_q(\alpha) = \frac{(1-q)^{(\alpha-1)}}{(1-q)^{\alpha-1}}, \quad \alpha \in \mathbb{R} - \{0, -1, -2, \dots\}. \tag{8}$$

Note that the q-gamma function satisfies $\Gamma_q(\alpha+1) = [\alpha]_q \Gamma_q(\alpha)$..

Definition 1.2 [13]: Let $\phi : J \to \mathbb{R}$ be a function and for $n \in \mathbb{N}$. The q-derivative of *order n of is defined by*

$$\left(D_q^0 \phi\right)(t) = \phi(t), \tag{9}$$

and

$$\left(D_q \phi\right)(t) = \left(D_q^1 \phi\right)(t) = \frac{\phi(t) - \phi(qt)}{(1-q)t}, \quad t \neq 0, \quad \left(D_q \phi\right)(0) = \lim_{t \to 0}\left(D_q \phi\right)(t), \tag{10}$$

$$\left(D_q^n \phi\right)(t) = \left(D_q^1 D_q^{n-1} \phi\right)(t), \quad t \in J, \quad n \in \mathbb{N}^*. \tag{11}$$

Set $J_t := \{tq^n : n \in \mathbb{N}\} \cup \{0\}$.

Definition 1.3 [13]: Let $\phi : J \to \mathbb{R}$ be a function. The q-integral is given by

$$\left(I_q \phi\right)(t) = \int_0^t \phi(s) d_q s = \sum_{n=0}^{\infty} t(1-q)q^n \phi(tq^n), \tag{12}$$

provided that the series converges.

Definition 1.4 [15]: The Riemann-Liouville's fractional q-integral of order $\alpha \in \mathbb{R}_+$ of a function $\phi : J \to \mathbb{R}$ is defined by

$$\begin{cases} \left(I_q^0 \phi\right)(x) = \phi(x); & \text{if } \alpha = 0, \\ \left(I_q^\alpha \phi\right)(x) = \dfrac{1}{\Gamma_q(\alpha)} \displaystyle\int_a^x (x - qt)^{(\alpha-1)} \phi(t) d_q t; & \text{if } \alpha > 0. \end{cases} \tag{13}$$

Note that for $\alpha = 1$, we have $\left(I_q^1 \phi\right)(t) = \left(I_q \phi\right)(t)$.

Lemma 1.5 [34]: For $\alpha \in \mathbb{R}_+$ and $\beta \in (-1, +\infty)$, we have

$$\left(I_q^\alpha (t-a)^{(\beta)}\right)(t) = \frac{\Gamma_q(\beta+1)}{\Gamma_q(\alpha+\beta+1)} (t-a)^{(\alpha+\beta)}, \quad 0 < a < t < T. \tag{14}$$

Particularly,

$$\left(I_q^\alpha 1\right)(t) = \frac{1}{\Gamma_q(\alpha+1)} t^{(\alpha)}. \tag{15}$$

Definition 1.6 [35]: Let $\phi : J \to \mathbb{R}$ be a function and for $\alpha \in \mathbb{R}_+$. The Riemann-Liouville's fractional q-derivative of order α is given by

$$\left({}^{RL}D_q^\alpha \phi\right)(x) = \begin{cases} \phi(x); & \text{if } \alpha = 0, \\ \left(D_q^{[\alpha]} I_q^{[\alpha]-\alpha} \phi\right)(x); & \text{if } \alpha > 0, \end{cases} \tag{16}$$

where $[\alpha]$ is the integer part of α.

Definition 1.7 [35]:Let $\phi : J \to \mathbb{R}$ be a function and for $\alpha \in \mathbb{R}_+$. The Caputo's fractional q-derivative of order α is given by

$$\left({}^{C}D_q^\alpha \phi\right)(x) = \begin{cases} \phi(x); & \text{if } \alpha = 0, \\ \left(I_q^{[\alpha]-\alpha} D_q^{[\alpha]} \phi\right)(x); & \text{if } \alpha > 0, \end{cases} \tag{17}$$

where $[\alpha]$ is the integer part of α.

Lemma 1.8 [35]: Let ϕ be a function defined on J and for $\alpha, \beta \in \mathbb{R}_+$. Then, the following identities hold:

i. $\left(I_q^\alpha I_q^\beta \phi\right)(t) = \left(I_q^{\alpha+\beta} \phi\right)(t)$.

ii. $\left(D_q^\alpha I_q^\alpha \phi\right)(t) = \phi(t)$.

Lemma 1.9 [35]: Let ϕ be a function defined on J and for $\alpha \in \mathbb{R}_+$. Then, the following equality holds:

$$\left(I_q^{\alpha C} D_q^\alpha \phi\right)(t) = \phi(t) - \sum_{k=0}^{[\alpha]-1} \frac{t^k}{\Gamma_q(k+1)} \left(D_q^k \phi\right)(0). \tag{18}$$

In particular, if $\alpha \in (0,1)$, then

$$\left(I_q^{\alpha C} D_q^{\alpha} \phi\right)(t) = \phi(t) - \phi(0). \tag{19}$$

Next, we present some fixed point theorems.

Theorem 1.10 [36] (Banach).

Let U be a non-empty closed subset of a Banach space $\mathbb{X}$ and $N : U \to U$ a contraction mapping, then the mapping N has a unique fixed point.

Theorem 1.11 [37] (Schaefer).

Let $\mathbb{X}$ be a Banach space and $N : \mathbb{X} \to \mathbb{X}$ be a completely continuous operator. If the set $\Omega(N) := \{y \in \mathbb{X} : y = \lambda N(y),\ \ for\ \ \lambda \in (0,1)\}$ is bounded, then the operator N has at least one fixed point.

Theorem 1.12 [36] (Nonlinear alternative of Leray-Schauder).

Let $\mathbb{X}$ be a Banach space and E be a closed, convex subset of $\mathbb{X}$. Let U be an open subset of E with $0 \in U$ and $N : \overline{U} \to E$ a continuous and compact operator. Then either

i. The operator N has fixed points, or

ii. There exist $y \in \partial U$ and $\lambda \in (0,1)$ with $y = \lambda N(y)$.

3. Main results

The purpose of this section is to provide results for the existence, uniqueness, and stability of solutions to the boundary value problem (2), by using the fixed point theorems and Ulam stability techniques.

Firstly, let us define what is meant by the integral solution to the boundary value problem (2).

Definition 1.13: A function $y \in C(J, \mathbb{R})$ is said to be a solution of the problem (2) if y satisfies the fractional q-difference equation $\left({}^C D_q^{\alpha} y\right)(t) = \phi(t, y(t))$ on J where $\alpha \in (1, 2]$, and satisfies the following integral boundary conditions:

$$y(0) - y'(0) = \int_0^T \varphi(s, y(s))ds \quad \text{and} \quad y(T) + y'(T) = \int_0^T \psi(s, y(s))ds.$$

Next, we need the following auxiliary lemma to determine the main results of the boundary value problem (2).

Lemma 1.14: Let $\sigma, \rho_1, \rho_2 \in C(J, \mathbb{R})$, the solution of the following boundary value problem:

$$\begin{cases} \left({}^C D_q^{\alpha} y\right)(t) = \sigma(t), t \in J = [0, T], \quad 1 < \alpha \leq 2, \\ \qquad y(0) - y'(0) = \int_0^T \rho_1(s)ds, \\ \qquad y(T) + y'(T) = \int_0^T \rho_2(s)ds, \end{cases} \tag{20}$$

is given by

$$y(t) = K(t) + \int_0^T H_q(t, s)\sigma(s)d_q s, \tag{21}$$

where

$$K(t) = \frac{(1+T-t)}{(2+T)} \int_0^T \rho_1(s)ds + \frac{(1+t)}{(2+T)} \int_0^T \rho_2(s)ds, \tag{22}$$

and

$$H_q(t,s) = \begin{cases} \dfrac{(t-qs)^{(\alpha-1)}}{\Gamma_q(\alpha)} - \dfrac{(1+t)(T-qs)^{(\alpha-1)}}{(2+T)\Gamma_q(\alpha)} - \dfrac{(1+t)(T-qs)^{(\alpha-2)}}{(2+T)\Gamma_q(\alpha-1)}; & 0 \le s < t, \\ -\dfrac{(1+t)(T-qs)^{(\alpha-1)}}{(2+T)\Gamma_q(\alpha)} - \dfrac{(1+t)(T-qs)^{(\alpha-2)}}{(2+T)\Gamma_q(\alpha-1)}; & t \le s \le T. \end{cases} \tag{23}$$

Proof: Let us start by applying the Riemann-Liouville's fractional q-integral of order $\alpha \in (1,2]$ to both sides of equation for the problem (20), and according to Lemma 9, we have

$$y(t) = \int_0^t \frac{(t-qs)^{(\alpha-1)}}{\Gamma_q(\alpha)} \sigma(s) d_q s + c_0 + c_1 t. \tag{24}$$

Next, we use the integral conditions of the problem (20), we obtain

$$c_0 - c_1 = \int_0^T \rho_1(s)ds, \tag{25}$$

and

$$c_0 + (1+T)c_1 + \int_0^T \frac{(T-qs)^{(\alpha-1)}}{\Gamma_q(\alpha)} \sigma(s) d_q s + \int_0^T \frac{(T-qs)^{(\alpha-2)}}{\Gamma_q(\alpha-1)} \sigma(s) d_q s = \int_0^T \rho_2(s)ds. \tag{26}$$

Eqs. (25) and (26) give

$$c_1 = \frac{1}{(2+T)} \left(\int_0^T \rho_2(s)ds - \int_0^T \rho_1(s)ds - \int_0^T \frac{(T-qs)^{(\alpha-1)}}{\Gamma_q(\alpha)} \sigma(s) d_q s - \int_0^T \frac{(T-qs)^{(\alpha-2)}}{\Gamma_q(\alpha-1)} \sigma(s) d_q s \right),$$

and

$$c_0 = \frac{(1+T)}{(2+T)} \int_0^T \rho_1(s)ds + \frac{1}{(2+T)} \left(\int_0^T \rho_2(s)ds - \int_0^T \frac{(T-qs)^{(\alpha-1)}}{\Gamma_q(\alpha)} \sigma(s) d_q s \right.$$
$$\left. - \int_0^T \frac{(T-qs)^{(\alpha-2)}}{\Gamma_q(\alpha-1)} \sigma(s) d_q s \right).$$

Then, replacing c_0 and c_1 in Eq. (24), we obtain

$$y(t) = \int_0^t \frac{(t-qs)^{(\alpha-1)}}{\Gamma_q(\alpha)}\sigma(s)d_qs + \frac{(1+T)}{(2+T)}\int_0^T \rho_1(s)ds + \frac{1}{(2+T)}\left(\int_0^T \rho_2(s)ds \right.$$
$$\left. - \int_0^T \frac{(T-qs)^{(\alpha-1)}}{\Gamma_q(\alpha)}\sigma(s)d_qs - \int_0^T \frac{(T-qs)^{(\alpha-2)}}{\Gamma_q(\alpha-1)}\sigma(s)d_qs\right) + \frac{t}{(2+T)}\left(\int_0^T \rho_2(s)ds \right.$$
$$\left. - \int_0^T \rho_1(s)ds - \int_0^T \frac{(T-qs)^{(\alpha-1)}}{\Gamma_q(\alpha)}\sigma(s)d_qs - \int_0^T \frac{(T-qs)^{(\alpha-2)}}{\Gamma_q(\alpha-1)}\sigma(s)d_qs\right). \tag{27}$$

Using the fact that $\int_0^T = \int_0^t + \int_t^T$, we get

$$y(t) = K(t) + \int_0^T H_q(t,s)\sigma(s)d_qs, \tag{28}$$

where

$$K(t) = \frac{(1+T-t)}{(2+T)}\int_0^T \rho_1(s)ds + \frac{(1+t)}{(2+T)}\int_0^T \rho_2(s)ds, \tag{29}$$

and

$$H_q(t,s) = \begin{cases} \dfrac{(t-qs)^{(\alpha-1)}}{\Gamma_q(\alpha)} - \dfrac{(1+t)(T-qs)^{(\alpha-1)}}{(2+T)\Gamma_q(\alpha)} - \dfrac{(1+t)(T-qs)^{(\alpha-2)}}{(2+T)\Gamma_q(\alpha-1)}; & 0 \le s < t, \\ -\dfrac{(1+t)(T-qs)^{(\alpha-1)}}{(2+T)\Gamma_q(\alpha)} - \dfrac{(1+t)(T-qs)^{(\alpha-2)}}{(2+T)\Gamma_q(\alpha-1)}; & t \le s \le T. \end{cases} \tag{30}$$

Finally, we obtain that y is a solution of the boundary value problem (20). The proof is completed.

3.1 Existence of solutions

In the following subsections, we prove existence and uniqueness results for the boundary value problem (2), by using a variety of fixed point theorems.

Let

$$H_q^* = \sup_{(t,s)\in J\times J} |H_q(t,s)|. \tag{31}$$

The first result is based on Banach's fixed point theorem (Theorem 1.15).

Theorem 1.15: Assume that the following hypotheses hold:

$(P1)$ The function $\phi : [0,T] \times \mathbb{R} \to \mathbb{R}$ is continuous.

$(P2)$ There exist non-negative constants L_ϕ, L_φ and L_ψ, such that for each $t \in J$ and each $x, y \in \mathbb{R}$, we have

$$|\phi(t,x)-\phi(t,y)|\leq L_{\phi}|x-y|.$$
$$|\varphi(t,x)-\varphi(t,y)|\leq L_{\varphi}|x-y|.$$
$$|\psi(t,x)-\psi(t,y)|\leq L_{\psi}|x-y|.$$

If

$$0<L_{\varphi}\frac{T(1+T)}{(2+T)}+L_{\psi}\frac{T(1+T)}{(2+T)}+L_{\phi}H_q^{*}T<1. \tag{32}$$

Then, the boundary value problem (2) has a unique solution on J.

Proof: To demonstrate this result, we first transform the problem (2) into a fixed point problem and define the operator

$$N:C(J,\mathbb{R})\to C(J,\mathbb{R})$$

Given by

$$(Ny)(t)=K(t)+\int_0^T H_q(t,s)\phi(s,y(s))d_qs, \tag{33}$$

where

$$K(t)=\frac{(1+T-t)}{(2+T)}\int_0^T\varphi(s,y(s))ds+\frac{(1+t)}{(2+T)}\int_0^T\psi(s,y(s))ds, \tag{34}$$

and the function $H_q(t,s)$ is given by Eq. (23), it's clear that the fixed points of the operator N are solutions of the problem (2).

Next, we will show that the operator N is contraction mapping on $C(J,\mathbb{R})$.

Let $x,y\in C(J,\mathbb{R})$ and for every $t\in J$, we have

$$\begin{aligned}|(Nx)(t)-(Ny)(t)|=\ &\left|\frac{(1+T-t)}{(2+T)}\int_0^T(\varphi(s,x(s))-\varphi(s,y(s)))ds\right.\\&+\frac{(1+t)}{(2+T)}\int_0^T(\psi(s,x(s))-\psi(s,y(s)))ds\\&\left.+\int_0^T H_q(t,s)(\phi(s,x(s))-\phi(sy(s)))d_qs\right|.\end{aligned}$$

Therefore,

$$\begin{aligned}|(Nx)(t)-(Ny)(t)|\leq\ &\frac{(1+T-t)}{(2+T)}\int_0^T|\varphi(s,x(s))-\varphi(s,y(s))|ds\\&+\frac{(1+t)}{(2+T)}\int_0^T|\psi(sx(s))-\psi(s,y(s))|ds\\&+\int_0^T|H_q(t,s)||\phi(s,x(s))-\phi(s,y(s))|d_qs.\end{aligned}$$

By hypothesis $(P2)$ and for each $t\in J$, we get

$$|(Nx)(t)-(Ny)(t)|\leq L_{\varphi}\frac{T(1+T)}{(2+T)}\|x-y\|_{\infty}+L_{\psi}\frac{T(1+T)}{(2+T)}\|x-y\|_{\infty}+L_{\phi}H_{q}^{*}T\|x-y\|_{\infty}.$$

Hence,

$$\|N(x)-N(y)\|_{\infty}\leq\left(L_{\varphi}\frac{T(1+T)}{(2+T)}+L_{\psi}\frac{T(1+T)}{(2+T)}+L_{\phi}H_{q}^{*}T\right)\|x-y\|_{\infty}. \tag{35}$$

By condition (32), the operator N is contraction, and according to Banach's fixed point theorem, we conclude that the operator N has a unique fixed point, which is the unique solution to the boundary value problem (2).

The second result depends on fixed point theorem of Schaefer (Theorem 1.11).

Theorem 1.16 Suppose that the hypothesis $(P1)$ is satisfied and the following hypothesis holds:

$(P3)$ There exist non-negative constants M_{ϕ}, M_{φ} and M_{ψ}, such that for each $t\in J$ and each $y\in\mathbb{R}$, we have

$$|\phi(t,y)|\leq M_{\phi}.$$
$$|\varphi(t,y)|\leq M_{\varphi}.$$
$$|\psi(t,y)|\leq M_{\psi}.$$

Then, the boundary value problem (2) has at least one solution on J.

Proof: In order to demonstrate this result, we will use the fixed point theorem of Schaefer, that is, we will prove that the operator N defined by Eq. (33) has a fixed point. So, the proof will be presented in four steps.

Step 1: *N is a continuous operator on $C(J,\mathbb{R})$.*

Let $\{y_n\}_{n\in\mathbb{N}}$ be a sequence such that $y_n\to y$ in $C(J,\mathbb{R})$. Then, for every $t\in J$, we have

$$\begin{aligned}|(Ny_n)(t)-(Ny)(t)|\leq\ &\frac{(1+T-t)}{(2+T)}\int_0^T|\varphi(s,y_n(s))-\varphi(s,y(s))|ds\\&+\frac{(1+t)}{(2+T)}\int_0^T|\psi(s,y_n(s))-\psi(s,y(s))|ds\\&+\int_0^T|H_q(t,s)||\phi(s,y_n(s))-\phi(s,y(s))|d_qs.\end{aligned}$$

Hence,

$$\begin{aligned}\|N(y_n)-N(y)\|_{\infty}\leq\ &\frac{T(1+T)}{(2+T)}\|\varphi(.,y_n(.))-\varphi(.,y(.))\|_{\infty}\\&+\frac{T(1+T)}{(2+T)}\|\psi(.,y_n(.))-\psi(.,y(.))\|_{\infty}\\&+H_q^{*}T\|\phi(.,y_n(.))-\phi(.,y(.))\|_{\infty}.\end{aligned}$$

Since $\phi,\varphi,$ and ψ are continuous functions, we find

$$\|N(y_n)-N(y)\|_{\infty}\to 0\quad as\quad n\to\infty. \tag{36}$$

Consequently, the operator N is continuous on $C(J,\mathbb{R})$.

Step 2: *N maps bounded sets into bounded sets in* $C(J,\mathbb{R})$.

Indeed, it is enough to show that for any $r>0$, there exists a positive constant R such that for each $y\in B_r=\{y\in C(J,\mathbb{R}):\|y\|_\infty\leq r\}$, we have $\|N(y)\|_\infty\leq R$.

Let $y\in B_r$. Then, for each $t\in J$, we have

$$\begin{aligned}|(Ny)(t)|\leq\ & \frac{(1+T-t)}{(2+T)}\int_0^T|\varphi(s,y(s))|ds+\frac{(1+t)}{(2+T)}\int_0^T|\psi(s,y(s))|ds\\&+\int_0^T|H_q(t,s)||\phi(s,y(s))|d_qs.\end{aligned}$$

By hypothesis $(P3)$ and for every $t\in J$, we get

$$|(Ny)(t)|\leq M_\varphi\frac{T(1+T)}{(2+T)}+M_\psi\frac{T(1+T)}{(2+T)}+M_\phi H_q^*T.$$

Hence,

$$\|N(y)\|_\infty\leq M_\varphi\frac{T(1+T)}{(2+T)}+M_\psi\frac{T(1+T)}{(2+T)}+M_\phi H_q^*T:=R. \tag{37}$$

Consequently, the operator N is uniformly bounded on B_r.

Step 3: *N maps bounded sets into equicontinuous sets of* $C(J,\mathbb{R})$.

Let $t_1,t_2\in J,\ \ t_1<t_2$ and let B_r be a bounded set of $C(J,\mathbb{R})$ as in Step 2. For $y\in B_r$ then, we have

$$\begin{aligned}|(Ny)(t_2)-(Ny)(t_1)|&=\left|\frac{(t_1-t_2)}{(2+T)}\int_0^T\varphi(s,y(s))ds+\frac{(t_2-t_1)}{(2+T)}\int_0^T\psi(s,y(s))ds\right.\\&\quad\left.+\int_0^T\big(H_q(t_2,s)-H_q(t_1,s)\big)\phi(s,y(s))d_qs\right|,\\&\leq\frac{(t_1-t_2)}{(2+T)}\int_0^T|\varphi(s,y(s))|ds+\frac{(t_2-t_1)}{(2+T)}\int_0^T|\psi(s,y(s))|ds\\&\quad+\int_0^T\big|H_q(t_2,s)-H_q(t_1,s)\big||\phi(s,y(s))|d_qs.\end{aligned}$$

By hypothesis $(P3)$, we get

$$|(Ny)(t_2)-(Ny)(t_1)|\leq M_\varphi\frac{T(t_1-t_2)}{(2+T)}+M_\psi\frac{T(t_2-t_1)}{(2+T)}+M_\phi T\sup_{s\in J}\big|H_q(t_2,s)-H_q(t_1,s)\big|. \tag{38}$$

As $t_1\to t_2$, the inequality above's right-hand side tends to zero.

As a result of Steps 1 to 3, thanks to the Arzela-Ascoli's theorem, we can deduce that the operator N is completely continuous.

Step 4: *A priori bound.*

Let us now show that the set $\Omega=\{y\in C(J,\mathbb{R}):y=\lambda N(y),0<\lambda<1\}$ is bounded.

Let $y\in\Omega$, then $y=\lambda N(y)$ for some $0<\lambda<1$. Thus, for every $t\in J$, we have

$$y(t)=\lambda\left(\frac{(1+T-t)}{(2+T)}\int_0^T\varphi(s,y(s))ds+\frac{(1+t)}{(2+T)}\int_0^T\psi(s,y(s))ds\qquad+\int_0^TH_q(t,s)\phi(s,y(s))d_qs\right).$$

Then, by hypothesis $(P3)$ (as in Step 2), it follows that, for each $t\in J$, we get

$$|y(t)|\leq M_{\varphi}\frac{T(1+T)}{(2+T)}+M_{\psi}\frac{T(1+T)}{(2+T)}+M_{\phi}H_{q}^{*}T. \tag{39}$$

Consequently, $\|y\|_{\infty}\leq R<\infty$, the set is thus bounded.

As a result of Steps 1 to 4 and according to the fixed point theorem of Shaefer, we conclude that the operator N has at least one fixed point which is the solution to the boundary value problem (2).

The third result is based on Leray-Schauder nonlinear alternative theorem (Theorem 1.12).

Theorem 1.17: Assume that the hypothesis $(P1)$ holds and the following hypotheses are satisfied:

$(P4)$ There exist $\Phi_{\phi},\Phi_{\varphi},\Phi_{\psi}\in L^{1}(J,\mathbb{R}_{+})$ and $\Psi_{\phi},\Psi_{\varphi},\Psi_{\psi}:[0,\infty)\rightarrow(0,\infty)$ are continuous and non-decreasing, such that for each $t\in J$ and each $y\in\mathbb{R}$, we have

$$|\phi(t,y)|\leq\Phi_{\phi}(t)\Psi_{\phi}(|y|).$$
$$|\varphi(t,y)|\leq\Phi_{\varphi}(t)\Psi_{\varphi}(|y|).$$
$$|\psi(t,y)|\leq\Phi_{\psi}(t)\Psi_{\psi}(|y|).$$

$(P5)$ There exists a number $\nu>0$ such that

$$\frac{\|\nu\|_{\infty}}{\frac{(1+T)}{(2+T)}\left(\|\Phi_{\varphi}\|_{L^{1}}\Psi_{\varphi}(\|\nu\|_{\infty})+\|\Phi_{\psi}\|_{L^{1}}\Psi_{\psi}(\|\nu\|_{\infty})\right)+H^{*}\|\Phi_{\phi}\|_{L^{1}}\Psi_{\phi}(\|\nu\|_{\infty})}>1. \tag{40}$$

Then, the boundary value problem (2) has at least one solution on J..

Proof: To establish that the operator N defined by eq. (33) has a fixed point, we will apply the nonlinear alternative of Leray-Schauder theorem. The operator N is completely continuous as shown in Theorem 1.16.

Let $y\in C(J,\mathbb{R})$ such that for each $t\in J$, we have $y(t)=\lambda(Ny)(t)$ for $\lambda\in(0,1)$. Then, from hypothesis $(P4)$ and for each $t\in J$, we obtain

$$\begin{aligned}|y(t)|\leq&\frac{(1+T-t)}{(2+T)}\int_{0}^{T}|\varphi(s,y(s))|ds+\frac{(1+t)}{(2+T)}\int_{0}^{T}|\psi(s,y(s))|ds\\&+\int_{0}^{T}|H_{q}(t,s)||\phi(s,y(s))|d_{q}s,\\\leq&\frac{(1+T-t)}{(2+T)}\int_{0}^{T}\Phi_{\varphi}(s)\Psi_{\varphi}(|y|)ds+\frac{(1+t)}{(2+T)}\int_{0}^{T}\Phi_{\psi}(s)\Psi_{\psi}(|y|)ds\\&+\int_{0}^{T}|H(t,s)|\Phi_{\phi}(s)\Psi_{\phi}(|y|)d_{q}s.\end{aligned}$$

Thus,

$$\|y\|_{\infty}\leq\frac{(1+T)}{(2+T)}\left(\|\Phi_{\varphi}\|_{L^{1}}\Psi_{\varphi}(\|y\|_{\infty})+\|\Phi_{\psi}\|_{L^{1}}\Psi_{\psi}(\|y\|_{\infty})\right)+H^{*}\|\Phi_{\phi}\|_{L^{1}}\Psi_{\phi}(\|y\|_{\infty}).$$

Hence,

$$\frac{\|y\|_\infty}{\frac{(1+T)}{(2+T)}\left(\|\Phi_\varphi\|_{L^1}\Psi_\varphi(\|y\|_\infty)+\|\Phi_\psi\|_{L^1}\Psi_\psi(\|y\|_\infty)\right)+H^*\|\Phi_\phi\|_{L^1}\Psi_\phi(\|y\|_\infty)}\leq 1. \tag{41}$$

Then, by hypothesis $(P5)$, there exists ν such that $\|y\|_\infty \neq \nu$. Let us set

$$U=\{y\in C(J,\mathbb{R}):\|y\|_\infty<\nu\}.$$

Clearly, the operator $N : \overline{U}\to C(J,\mathbb{R})$ is completely continuous. From the choice of U, there is no $y\in\partial U$ such that $y=\lambda N(y)$, for some $\lambda\in(0,1)$. Therefore, according to the theorem of Leray-Schauder nonlinear alternative, we conclude that the operator N has at least one fixed point $y\in\overline{U}$, which is a solution of the boundary value problem (2). The proof is completed.

3.2 Ulam stability results

This part focuses on defining and studying different types of Ulam stability for the boundary value problem (2). Based on the references [17, 30–33], we provide the following definitions:

Definition 1.18: The problem (2) is Ulam-Hyers stable if there exists a real number $C>0$ such that for each $\varepsilon>0$ and for each solution $x\in C(J,\mathbb{R})$ of the following inequality:

$$|\left({}^CD_q^\alpha x\right)(t)-\phi(t,x(t))|\leq\varepsilon,\quad t\in J, \tag{42}$$

there exists a solution $y\in C(J,\mathbb{R})$ of the problem (2) with

$$\|x-y\|_\infty\leq C\varepsilon. \tag{43}$$

Definition 1.19: The problem (2) is generalized Ulam-Hyers stable if there exists $\theta\in C(\mathbb{R}_+,\mathbb{R}_+)$ with $\theta(0)=0$, such that for each $\varepsilon>0$, and for each solution $x\in C(J,\mathbb{R})$ of the inequality (42), there exists a solution $y\in C(J,\mathbb{R})$ of the problem (2) with

$$\|x-y\|_\infty\leq\theta(\varepsilon). \tag{44}$$

Definition 1.20: The problem (2) is Ulam-Hyers-Rassias stable with respect to χ if there exists $C_\chi>0$ such that for each $\varepsilon>0$ and for each solution $x\in C(J,\mathbb{R})$ of the following inequality:

$$|\left({}^CD_q^\alpha x\right)(t)-\phi(t,x(t))|\leq\varepsilon\chi(t),\quad t\in J, \tag{45}$$

there exists a solution $y\in C(J,\mathbb{R})$ of the problem (2) with

$$\|x-y\|_\infty\leq C_\chi\varepsilon\chi(t),\quad t\in J. \tag{46}$$

Remark 1 Clearly: Definition 18 $\Rightarrow$ Definition 1.19.

Theorem 1.21: Assume that the hypotheses $(P1)-(P2)$ and condition (32) are satisfied. Then, the boundary value problem (2) is Ulam-Hyers stable.

Proof: Let $x \in C(J, \mathbb{R})$ be a solution of the inequality (42) and let $y \in C(J, \mathbb{R})$ be the unique solution of the problem (2), we have

$$y(t) = K(t) + \int_0^T H_q(t,s)\phi(s,y(s))d_qs, \tag{47}$$

where

$$K(t) = \frac{(1+T-t)}{(2+T)}\int_0^T \varphi(s,y(s))ds + \frac{(1+t)}{(2+T)}\int_0^T \psi(s,y(s))ds, \tag{48}$$

and the function $H_q(t,s)$ is given by eq. (23).

From the inequality (42) and for each $t \in J$, we obtain

$$\left| \begin{array}{c} x(t) - \dfrac{(1+T-t)}{(2+T)}\displaystyle\int_0^T \varphi(s,x(s))ds \\ -\dfrac{(1+t)}{(2+T)}\displaystyle\int_0^T \psi(s,x(s))ds - \displaystyle\int_0^T H_q(t,s)\phi(s,x(s))d_qs \end{array} \right| \leq I_q^\alpha \varepsilon,$$

$$\leq \frac{t^{(\alpha)}}{\Gamma_q(\alpha+1)}\varepsilon.$$

Then, for each $t \in J$, we can write

$$\begin{aligned} |x(t)-y(t)| \leq & \left| x(t) - \frac{(1+T-t)}{(2+T)}\int_0^T \varphi(s,y(s))ds - \frac{(1+t)}{(2+T)}\int_0^T \psi(s,y(s))ds, \right. \\ & \left. -\int_0^T H_q(t,s)\phi(s,y(s))d_qs, \right|, \\ \leq & \left| , x(t) - \frac{(1+T-t)}{(2+T)}\int_0^T \varphi(s,x(s))ds - \frac{(1+t)}{(2+T)}\int_0^T \psi(s,x(s))ds \right. \\ & -\int_0^T H_q(t,s)\phi(s,x(s))d_qs + \frac{(1+T-t)}{(2+T)}\int_0^T (\varphi(s,x(s)) - \varphi(s,y(s)))ds \\ & \left. , + \frac{(1+t)}{(2+T)}\int_0^T (\psi(sx(s)) - \psi(s,y(s)))ds + \int_0^T H_q(t,s)(\phi(s,x(s)) - \phi(s,y(s)))d_qs \right|, \\ \leq & \left| x(t) - \frac{(1+T-t)}{(2+T)}\int_0^T \varphi(s,x(s))ds - \frac{(1+t)}{(2+T)}\int_0^T \psi(s,x(s))ds \right. \\ & -\int_0^T H_q(t,s)\phi(s,x(s))d_qs| + |\frac{(1+T-t)}{(2+T)}\int_0^T (\varphi(s,x(s)) - \varphi(s,y(s)))ds \\ & \left. , + \frac{(1+t)}{(2+T)}\int_0^T \Big(\psi(s,x(s)) - \psi(s,y(s)))ds + \int_0^T H_q(t,s)(\phi(s,x(s)) - \phi(sy(s)))d_qs, \right|. \end{aligned}$$

Hence,

$$\begin{aligned} |x(t)-y(t)| \leq & \frac{t^{(\alpha)}}{\Gamma_q(\alpha+1)}\varepsilon + \frac{(1+T-t)}{(2+T)}\int_0^T |\varphi(s,x(s)) - \varphi(s,y(s))|ds + \frac{(1+t)}{(2+T)} \\ & \times \int_0^T \Big| \psi(s,x(s)) - \psi(s,y(s))|ds + \int_0^T H_q(t,s)\Big|\phi(s,x(s)) - \phi(s,y(s))|d_qs. \end{aligned}$$

Using the hypothesis $(P2)$ and for every $t \in J$, we get

$$\|x-y\|_\infty \le \frac{T^{(\alpha)}}{\Gamma_q(\alpha+1)}\varepsilon+\left(L_\varphi\frac{T(1+T)}{(2+T)}+L_\psi\frac{T(1+T)}{(2+T)}+L_\phi H_q^* T\right)\|x-y\|_\infty.$$

So

$$\|x-y\|_\infty \le \frac{\dfrac{T^{(\alpha)}}{\Gamma_q(\alpha+1)}}{1-\left(L_\varphi\dfrac{T(1+T)}{(2+T)}+L_\psi\dfrac{T(1+T)}{(2+T)}+L_\phi H_q^* T\right)}\varepsilon, \qquad (49)$$
$$:= C\varepsilon.$$

Consequently, the boundary value problem (2) is Ulam-Hyers stable.

Taking $\theta(\varepsilon)=C\varepsilon$; $\theta(0)=0$, we can conclude that the boundary value problem (2) is generalized Ulam-Hyers stable.

Theorem 1.22: Suppose that the hypotheses $(P1)-(P2)$ and condition (32) are satisfied and the following hypothesis holds:

$(P6)$ Let $\chi\in C(J,R_+)$ be an increasing function. There exists $\lambda_\chi>0$, such that for every $t\in J$, we have

$$I_q^\alpha\chi(t)\le\lambda_\chi\chi(t). \qquad (50)$$

Then, the boundary value problem (2) is Ulam-Hyers-Rassias stable.

Proof: Let $x\in C(J,\mathbb{R})$ be a solution of the inequality (45) and let $y\in C(J,\mathbb{R})$ be the unique solution of the problem (2), we have

$$y(t)=K(t)+\int_0^T H_q(t,s)\phi(s,y(s))d_qs, \qquad (51)$$

where

$$K(t)=\frac{(1+T-t)}{(2+T)}\int_0^T\varphi(s,y(s))ds+\frac{(1+t)}{(2+T)}\int_0^T\psi(s,y(s))ds, \qquad (52)$$

and the function $H_q(t,s)$ is given by eq. (23).

From the inequality (45), hypothesis $(P6)$ and for each $t\in J$, we obtain

$$\left|\begin{array}{c} x(t)-\dfrac{(1+T-t)}{(2+T)}\displaystyle\int_0^T\varphi(s,x(s))ds \\ -\dfrac{(1+t)}{(2+T)}\displaystyle\int_0^T\psi(s,x(s))ds-\int_0^T H_q(t,s)\phi(s,x(s))d_qs\end{array}\right| \le \varepsilon I_q^\alpha\chi(t),$$
$$\le\varepsilon\lambda_\chi\chi(t).$$

Then, for each $t\in J$, we can write

$$\begin{aligned}
|x(t)-y(t)| \leq & \left|x(t)-\frac{(1+T-t)}{(2+T)}\int_0^T \varphi(s,y(s))ds-\frac{(1+t)}{(2+T)}\int_0^T \psi(s,y(s))ds\right.\\
& \left.-\int_0^T H_q(t,s)\phi(s,y(s))d_qs\right|,\\
\leq & \left|x(t)-\frac{(1+T-t)}{(2+T)}\int_0^T \varphi(s,x(s))ds-\frac{(1+t)}{(2+T)}\int_0^T \psi(s,x(s))ds\right.\\
& -\int_0^T H_q(t,s)\phi(s,x(s))d_qs\frac{(1+T-t)}{(2+T)}\int_0^T(\varphi(s,x(s))-\varphi(s,y(s)))ds\\
& \left.+\frac{(1+t)}{(2+T)}\int_0^T(\psi(s,x(s))-\psi(s,y(s)))ds+\int_0^T H_q(t,s)(\phi(s,x(s))-\phi(s,y(s)))d_qs\right|,\\
\leq & \left|x(t)-\frac{(1+T-t)}{(2+T)}\int_0^T \varphi(s,x(s))ds-\frac{(1+t)}{(2+T)}\int_0^T \psi(s,x(s))ds\right.\\
& -\int_0^T H_q(t,s)\phi(s,x(s))d_qs|+|\frac{(1+T-t)}{(2+T)}\int_0^T(\varphi(s,x(s))-\varphi(s,y(s)))ds\\
& \left.+\frac{(1+t)}{(2+T)}\int_0^T\Big(\psi(s,x(s))-\psi(s,y(s)))ds+\int_0^T H_q(t,s)\big(\phi(s,x(s))-\phi(s,y(s))\big)d_qs\right|.
\end{aligned}$$

Thus,

$$\begin{aligned}
|x(t)-y(t)| \leq & \ \varepsilon\lambda_\chi\chi(t)+\frac{(1+T-t)}{(2+T)}\int_0^T|\varphi(s,x(s))-\varphi(s,y(s))|ds+\frac{(1+t)}{(2+T)}\\
& \times\int_0^T\Big|\psi(s,x(s))-\psi(s,y(s))|ds+\int_0^T H_q(t,s)\big|\phi(s,x(s))-\phi(s,y(s))|d_qs.
\end{aligned}$$

Using the hypothesis $(P2)$ and for each $t\in J$, we find

$$\|x-y\|_\infty \leq \varepsilon\lambda_\chi\chi(t)+\left(L_\varphi\frac{T(1+T)}{(2+T)}+L_\psi\frac{T(1+T)}{(2+T)}+L_\phi H_q^*T\right)\|x-y\|_\infty.$$

So

$$\begin{aligned}
\|x-y\|_\infty \leq & \frac{\varepsilon\lambda_\chi\chi(t)}{1-\left(L_\varphi\frac{T(1+T)}{(2+T)}+L_\psi\frac{T(1+T)}{(2+T)}+L_\phi H_q^*T\right)},\\
:= & C_\chi\varepsilon\chi(t).
\end{aligned} \tag{53}$$

Consequently, the boundary value problem (2) is Ulam-Hyers-Rassias stable.

4. Applications

This section includes two examples that highlight our main results.

Example 4.1: Consider the following boundary value problem of fractional q-difference equations:

$$\begin{cases} \left({}^{C}D_{\frac{1}{2}}^{\frac{4}{3}}y\right)(t) = \dfrac{t^2|y(t)|}{6e^t(1+|y(t)|)}, \ t \in J = [0,1], \ 1 < \alpha \leq 2, \\ y(0) - y'(0) = \displaystyle\int_0^1 \frac{s^2+1}{3}|y(s)|ds, \\ y(1) + y'(1) = \displaystyle\int_0^1 \frac{s^2-1}{5}|y(s)|ds, \end{cases} \tag{54}$$

where $\alpha = \frac{4}{3}$, $q = \frac{1}{2}, T = 1$, and for each $(t,y) \in J \times \mathbb{R}_+$, we have

$$\phi(t,y) = \frac{t^2 y}{6e^t(1+y)}, \quad \varphi(t,y) = \frac{t^2+1}{3}y, \quad \psi(t,y) = \frac{t^2-1}{5}y.$$

Clearly, the functions ϕ, φ, and ψ are continuous.
Let $x, y \in \mathbb{R}_+$ and for each $t \in J$. Then, we have

$$\begin{aligned} |\phi(t,x) - \phi(t,y)| &= \left|\frac{t^2}{6e^t}\left(\frac{x}{1+x} - \frac{y}{1+y}\right)\right|, \\ &\leq \frac{t^2}{6e^t}|x-y|, \\ &\leq \frac{1}{6}|x-y|, \end{aligned}$$

and

$$\begin{aligned} |\varphi(t,x) - \varphi(t,y)| &\leq \frac{1}{3}|x-y|, \\ |\psi(t,x) - \psi(t,y)| &\leq \frac{1}{5}|x-y|. \end{aligned}$$

From Eq. (23), we have

$$H_q^* = \sup_{(t,s) \in J \times J} |H_q(t,s)| = \frac{5}{3\Gamma_{\frac{1}{2}}\left(\frac{4}{3}\right)} + \frac{2}{3\Gamma_{\frac{1}{2}}\left(\frac{1}{3}\right)}. \tag{55}$$

Hence, the hypothesis $(P2)$ holds with $L_\phi = \frac{1}{6}$, $L_\varphi = \frac{1}{3}$, and $L_\psi = \frac{1}{5}$. We shall check that the condition (32) is satisfied with $T = 1$. Indeed,

$$\begin{aligned} \left(L_\varphi \frac{T(1+T)}{(2+T)} + L_\psi \frac{T(1+T)}{(2+T)} + L_\phi H_q^* T\right) &= \left(\frac{2}{9} + \frac{2}{15} + \frac{1}{6}\left(\frac{5}{3\Gamma_{\frac{1}{2}}\left(\frac{4}{3}\right)} + \frac{2}{3\Gamma_{\frac{1}{2}}\left(\frac{1}{3}\right)}\right)\right), \\ &= 0.7057 < 1. \end{aligned}$$

Then, by Theorem 1.15, the problem (54) has a unique solution on $[0,1]$ and from Theorem 1.21, the problem (54) is Ulam-Hyers stable.

Now, let $\chi(t) = t^2$ for each $t \in J$, we have

$$I_{\frac{1}{2}}^{\frac{4}{3}}\chi(t) = \frac{\Gamma_{\frac{1}{2}}(3)}{\Gamma_{\frac{1}{2}}\left(\frac{13}{3}\right)} t^{2+\frac{3}{4}} \leq \frac{3}{2\Gamma_{\frac{1}{2}}\left(\frac{13}{3}\right)} t^2 = \lambda_\chi \chi(t). \tag{56}$$

Thus, the hypothesis $(P6)$ is satisfied with $\chi(t) = t^2$ and $\lambda_\chi = \frac{3}{2\Gamma_{\frac{1}{2}}\left(\frac{13}{3}\right)}$. Then, all the hypotheses of Theorem 1.22 hold. So, the problem (54) is Ulam-Hyers-Rassias stable.

On the other hand, let $y \subset \mathbb{R}_+$ and for every $t \subset J$, we have

$$|\phi(t,y)| \leq \frac{1}{6}, \quad |\varphi(t,y)| \leq \frac{1}{3} \quad and \quad |\psi(t,y)| \leq \frac{1}{5}.$$

Hence, the hypothesis (P3) holds with $M_\phi = \frac{1}{6}$, $M_\varphi = \frac{1}{3}$ and $M_\psi = \frac{1}{5}$.

Therefore, according to Theorem 1.16, the problem (54) has at least one solution on $[0,1]$.

Example 4.2: Consider the following boundary value problem of fractional q-difference equations

$$\begin{cases} \left({}^C D_{\frac{1}{7}}^{\frac{8}{7}} y\right)(t) = \dfrac{te^{-t}}{t+6}\cos(y(t)), \quad t \in J = [0,1], \quad 1 < \alpha \leq 2, \\ y(0) - y'(0) = \displaystyle\int_0^1 \frac{s}{2}\cos(y(s))ds, \\ y(1) + y'(1) = \displaystyle\int_0^1 \frac{s}{2}\sin(y(s))ds, \end{cases} \tag{57}$$

where $\alpha = \frac{8}{7}$, $q = \frac{1}{7}$, $T = 1$, and for each $(t,y) \in J \times \mathbb{R}$, we have

$$\phi(t,y) = \frac{te^{-t}}{t+6}\cos(y), \quad \varphi(t,y) = \frac{t}{2}\cos(y), \quad \psi(t,y) = \frac{t}{2}\sin(y).$$

Clearly, the functions $\phi, \varphi,$ and ψ is continuous.

Let $x, y \in \mathbb{R}$ and for $t \in J$. Then, we have

$$\begin{aligned} |\phi(t,x) - \phi(t,y)| &= \left|\frac{te^{-t}}{t+6}(\cos(x) - \cos(y))\right|, \\ &\leq \frac{te^{-t}}{t+6}|\cos(x) - \cos(y)|, \\ &\leq \frac{1}{6}|x-y|, \end{aligned}$$

and

$$\begin{aligned} |\varphi(t,x) - \varphi(t,y)| &\leq \frac{1}{2}|x-y|, \\ |\psi(t,x) - \psi(t,y)| &\leq \frac{1}{2}|x-y|. \end{aligned}$$

From Eq. (23), we have

$$H_q^* = \sup_{(t,s)\in J\times J} |H_q(t,s)| = \frac{5}{3\Gamma_{\frac{1}{7}}\left(\frac{8}{7}\right)} + \frac{2}{3\Gamma_{\frac{1}{7}}\left(\frac{1}{7}\right)}. \tag{58}$$

Hence, the hypothesis (P2) holds with $L_\phi = \frac{1}{6}$, $L_\varphi = \frac{1}{2}$, and $L_\psi = \frac{1}{2}$. We shall check that the condition (32) is satisfied with $T = 1$. Indeed

$$\begin{aligned}\left(L_\varphi \frac{T(1+T)}{(2+T)} + L_\psi \frac{T(1+T)}{(2+T)} + L_\phi H_q^* T\right) &= \left(\frac{1}{3} + \frac{1}{3} + \frac{1}{6}\left(\frac{5}{3\Gamma_{\frac{1}{7}}\left(\frac{8}{7}\right)} + \frac{2}{3\Gamma_{\frac{1}{7}}\left(\frac{1}{7}\right)}\right)\right), \\ &= 0.9833 < 1.\end{aligned}$$

Then, by Theorem 1.15, the problem (57) has a unique solution on $[0,1]$ and from Theorem 1.21, the problem (57) is Ulam-Hyers stable.

Now, let $\chi(t) = t^2$ for each $t \in J$, we have

$$I_{\frac{1}{7}}^{\frac{4}{3}}\chi(t) = \frac{\Gamma_{\frac{1}{7}}(3)}{\Gamma_{\frac{1}{7}}\left(\frac{29}{7}\right)} t^{2+\frac{8}{7}} \leq \frac{\Gamma_{\frac{1}{7}}(3)}{\Gamma_{\frac{1}{7}}\left(\frac{29}{7}\right)} t^2 = \lambda_\chi \chi(t). \tag{59}$$

Thus, the hypothesis $(P6)$ is satisfied with $\chi(t) = t^2$ and $\lambda_\chi = \frac{\Gamma_{\frac{1}{7}}(3)}{\Gamma_{\frac{1}{7}}\left(\frac{29}{7}\right)}$. Hence, all the hypotheses of Theorem 1.22 hold. So, the problem (57) is Ulam-Hyers-Rassias stable.

On the other hand, let $y \in \mathbb{R}_+$ and for every $t \in J$, we have

$$|\phi(t,y)| \leq \frac{1}{6}, \quad |\varphi(t,y)| \leq \frac{1}{2} \quad \textit{and} \quad |\psi(t,y)| \leq \frac{1}{2}.$$

Then, the hypothesis (P3) holds with $M_\phi = \frac{1}{6}$, $M_\varphi = \frac{1}{2}$, and $M_\psi = \frac{1}{2}$.

Therefore, all the hypotheses of Theorem 1.16 are satisfied. So, the problem (57) has at least one solution on $[0,1]$.

5. Conclusions

In this work, we have offered the existence and uniqueness of solutions to the boundary value problem for fractional q-difference equations and integral conditions. The uniqueness result was obtained by applying Banach's fixed point theorem, while the existence results are proved *via* Schaefer's fixed point theorem and Leray-Schauder nonlinear alternative. Further, we obtained some results for Ulam-Hyers stability and Ulam-Hyers-Rassias stability. Finally, we illustrated our main results by providing examples.

Author details

Nadia Allouch*† and Samira Hamani†
Laboratory of Pure and Applied Mathematics, University of Mostaganem, Mostaganem, Algeria

*Address all correspondence to: nadia.allouch.etu@univ-mosta.dz

† These authors contributed equally.

IntechOpen

© 2024 The Author(s). Licensee IntechOpen. This chapter is distributed under the terms of the Creative Commons Attribution License (http://creativecommons.org/licenses/by/4.0), which permits unrestricted use, distribution, and reproduction in any medium, provided the original work is properly cited.

References

[1] Samko SG, Kilbas AA, Marichev OI. Fractional Integrals and Derivatives. Theory and Applications. Yverdon: Gordon and Breach; 1993

[2] Miller KS, Ross B. An Introduction to the Fractional Calculus and Differential Equations. New York: John Wiley; 1993

[3] Podlubny I. Fractional differential equations. San Diego: Academic Press; 1998

[4] Hilfer R. Applications of Fractional Calculus in Physics. Singapore: World Scientific; 2000

[5] Kilbas AA, Srivastava HM, Trujillo JJ. Theory and Applications of Fractional Differential Equations. North-Holland Mathematical Studies. Amsterdam: Elsevier Science BV; 2006. p. 204

[6] Tarasov VE. Fractional Dynamics : Application of Fractional Calculus to Dynamics of Particles. Heidelberg, Higher Education Press, Beijing: Springer; 2010

[7] Agarwal RP, Benchohra M, Hamani S. A survey on existence results for boundary value problems for nonlinear fractional differential equations and inclusions. Acta Applicandae Mathematicae. 2010;**109**(3):973-1033

[8] Benchohra M, Graef JR, Hamani S. Existence results for boundary value problems with nonlinear fractional differential equations. Applicable Analysis. 2008;**87**:851-863

[9] Afshari H, Karapinar E. Solution of the fractional differential equations in the setting of b-metric space. Carpathian Mathematical Publications. 2021;**3**:764-774

[10] Jackson F. On $_q$-definite integrals. The Quarterly Journal of Pure and Applied Mathematics. 1910;**41**:193-203

[11] Jackson F. On $_q$-functions and a certain difference operator. Transactions of the Royal Society of Edinburgh. 1908; **46**:253-281

[12] Gasper G, Rahman M. Basic Hypergeometric Series. Cambridge: Cambridge University Press; 1990

[13] Kac V, Cheung P. Quantum calculus. New York: Springer; 2002

[14] Al-Salam W. Some fractional $_q$-integrals and $_q$-derivatives. Proceedings of the Edinburgh Mathematical Society. 1966–1967;**15**(2, 140):135

[15] Agarwal R. Certain fractional $_q$-integrals and $_q$-derivatives. Proceedings of the Cambridge Philosophical Society. 1969

[16] Abbas S, Benchohra M, Henderson J. Oscillation for coupled fractional $_q$-difference systems. Fractional Calculus and Applied. 2021;**12**(1):143-155

[17] Abbas S, Benchohra M, Laledj N, Zhou Y. Exictence and Ulam stability for implicit fractional q-difference equation. Advances in Difference Equations. 2019; **2019**(480):1-12

[18] Laledj N, Salim A, Lazreg JE, Abbas S, Ahmad B, Benchohra M. On implicit fractional $_q$-difference equations : Analysis and stability. Mathematicsl Methods in the Applied Sciences. 2022:1-23

[19] Ahmad B, Ntouyas SK, Purnaras IK. Existence results for nonlocal boundary value problems of nonlinear fractional $_q$-difference equations. Advances in Difference Equations. 2012;**2011**(140): 1-15

[20] Ahmad B. Boundary value problem for nonlinear third order $_q$-difference

equations. Electronic Journal of Differential Equations. 2011;**2011**(94):1-7

[21] Allouch N, Hamani S. Boundary value problem for fractional q-difference equations in Banach space. The Rocky Mountain Journal of Mathematics. 2023; **53**(4):1001-1010

[22] Allouch N, Hamani S, Graef JR. Boundary value problem for fractional q-difference equations with integral condition in Banach space. Fractal and Fractional. 2022;**6**(94):1-11

[23] Allouch N, Hamani S, Henderson J. Boundary value problem for fractional q-difference equations. Nonlinear Dynamics and Systems Theory. 2024; **24**(2):111-122

[24] Annaby MH, Mansour ZS. q-Fractional Calculus and Equations. Heidelberg: Lecture Notes in Mathematics. Springer; 2012. p. 2056

[25] Hyers DH. On the stability of the linear functional equation. Proceedings of the National Academy of Sciences of the USA. 1941;**27**(4):222-224

[26] Ulam SM. A Collection of Mathematical Problems. New York: Interscience Publishers. Inc; 1960

[27] Rassias TM. On the stability of linear mappings in Banach spaces. Proceedings of American Mathematical Society. 1978; **72**:297-300

[28] Abbas S, Benchohra M, Graef JR, Henderson J. Implicit Fractional Differential and Integral Equations (Existence and Stability). Berlin: Gruyter; 2018

[29] Jung SM. Hyers-Ulam-Rassias Stability of Functional Equations in Nonlinear Analysis. New York: Springer; 2011. p. 204

[30] Rus IA. Remarks on Ulam stability of the operatorial equations. Fixed Point Theory. 2009;**10**:305-320

[31] Rus IA. Ulam stability of ordinary differential equations. Studia Universitatis Babes-Bolyai, Math LIV. 2009;**4**:125-133

[32] Taieb A, Dahmani Z. Ulam-Hyers-rassias stability of fractional lane-emden equations. ROMAI Journal. 2019;**15**(1): 133-153

[33] Houas M, Martnez F, Samei ME, Kaabar MKA. Uniqueness and Ulam-Hyers-Rassias stability results for sequential fractional pantograph q-differential equations. Journal of Inequalities and Applications. 2022; **2022**(93):1-24

[34] Rajkovic PM, Marinkovic SD, Stankovic MS. Fractional integrals and derivatives in q-calculus. Applicable Analysis and Discrete Mathematics. 2007;**1**:311-323

[35] Rajkovic PM, Marinkovic SD, Stankovic MS. On q-analogues of Caputo derivative and Mittag-Leffler function. Fractional Calculus and Applied Analysis. 2007;**10**(4):359-373

[36] Granas A, Dugundji J. Fixed Point Theory. New York: Springer; 2003

[37] Smart DR. Fixed Point Theorems. Cambridge University Press; 1980

Chapter 4

Entropy Solution to a Singular Fractional Problem

Masoud Bayrami and Mahmoud Hesaaraki

Abstract

The goal is to provide a brief overview of the existence results for the following singular partial differential equation governed by the fractional Laplacian operator, involving singular nonlinearities and irregular data. The model problem we consider is as follows:

$$\begin{cases} (-\Delta)^s v = \dfrac{\mu}{v^\gamma} + \nu & \text{in } \Omega, \\ v > 0 & \text{in } \Omega, \\ v = 0 & \text{in } \Omega^c, \end{cases}$$

for $0 < s < 1$, $\gamma > 0$, and $\Omega \subset \mathbb{R}^N$ (with $N > 2s$) a bounded smooth domain, where $\Omega^c := (\mathbb{R}^N \backslash \Omega)$. Moreover, $0 \leq \mu, \nu \in L^1(\Omega)$, or more generally, non-negative bounded Radon measures. For any $\gamma > 0$, we will establish the existence of a positive weak solution to the aforementioned problem. Additionally, when $0 < \gamma \leq 1$, for certain specific data μ and ν, we will establish both the existence and uniqueness of a distinct solution type known as an entropy solution. We will also investigate the uniqueness of the weak solution for $\gamma > 1$ and show the equivalence between entropy and weak solutions when $0 < \gamma \leq 1$. Lastly, we will explore the possibility of relaxing certain assumptions on the data.

Keywords: fractional Laplacian equation, singular nonlinearities, irregular data, positive entropy solution, positive solution, entropy solution, uniqueness

1. Introduction

We explore the presence of a positive solution, considering various perspectives on the concept of a solution, for the following singular elliptic problem:

$$\begin{cases} (-\Delta)^s u = \dfrac{\mu}{u^\gamma} + \nu & \text{in } \Omega, \\ u > 0 & \text{in } \Omega, \\ u = 0 & \text{in } \Omega^c := (\mathbb{R}^N \backslash \Omega). \end{cases} \tag{1}$$

Here, $\Omega \subset \mathbb{R}^N$ $(N > 2s)$ represents an open, bounded domain with a smooth boundary, where $s \in (0, 1)$, $\gamma > 0$, and μ, ν are irregular data; for instance, they may be non-negative functions in $L^1(\Omega)$ or, more broadly, non-negative bounded Radon measures.

In their work [1], Lazer and McKenna showed that if Ω is a bounded domain with a smooth boundary, $p(x)$ is smooth on $\overline{\Omega}$, and α is any positive constant, then there exists a unique solution in $C^2(\Omega)$, which is continuous up-to-boundary of Ω, to the following problem:

$$\begin{cases} -\Delta u = \dfrac{p(x)}{u^\alpha} & \text{in } \Omega, \\ u > 0 & \text{in } \Omega, \\ u = 0 & \text{on } \partial\Omega. \end{cases} \tag{2}$$

However, this solution does not lie in $C^2(\overline{\Omega})$. Moreover, Lazer and McKenna identified a necessary and sufficient condition on α for the solution to possess a finite Dirichlet integral. Specifically, they proved that $u \in H_0^1(\Omega)$ if and only if $\alpha < 3$. Additionally, they demonstrated that $u \notin C^1(\overline{\Omega})$ when $\alpha > 1$. In Ref. [2], Bougherara, Giacomoni, and Hernandez demonstrated the existence of solutions to this problem for any $\alpha \in \mathbb{R}$, where $p \in C(\Omega)$ exhibits behavior similar to $\text{dist}(x, \partial\Omega)^{-\beta}$ near the boundary, with $0 \leq \beta < 2$. They investigated the behavior of solutions, providing a gradient estimate near the boundary. Additionally, they demonstrated that the solution belongs to $W_0^{1,q}(\Omega)$ for $1 < q < \frac{1+\alpha}{\alpha+\beta-1}$.

Our results in this note follow a similar line of research, focusing on the non-local case. Problems of this form have been widely studied for both their mathematical significance [1–29] and their relevance to physical phenomena, such as in the theory of pseudoplastic fluids [30]. We encourage interested readers to consult the recent survey article [31] and the references therein for further insights on this topic when $s = 1$.

In this note, we aim to provide a concise overview of the weak formulation of a solution to problem (1) in the context of the non-local operator $(-\Delta)^s$, where $0 < s < 1$. This operator, referred to as the restricted fractional Laplacian, is a non-local extension of the classical differential operator, which is inherently local in nature:

$$-\Delta u(x) = -\sum_{i=1}^{N} \frac{\partial^2 u}{\partial x_i^2}(x). \tag{3}$$

It is defined *via* a singular integral operator as follows:

$$(-\Delta)^s u(x) = C_{N,s} \lim_{\varepsilon \to 0^+} \int_{|x-y| \geq \varepsilon} \frac{u(x) - u(y)}{|x - y|^{N+2s}} dy, \qquad u \in \mathcal{S}(\mathbb{R}^N), \tag{4}$$

where $\mathcal{S}(\mathbb{R}^N)$ denotes the Schwartz space on $\mathbb{R}^N$, and $C_{N,s} = \frac{4^s \Gamma\left(\frac{N}{2}+s\right)}{\pi^{\frac{N}{2}} |\Gamma(-s)|}$ is a constant (Γ representing the Gamma function) such that the following holds:

$$(-\Delta)^s u = \mathcal{F}^{-1}\left(|\xi|^{2s} \hat{u}(\xi)\right), \tag{5}$$

where $\mathcal{F}u = \hat{u}$ denotes the Fourier transform of u. By limiting the fractional Laplacian operator to act exclusively on smooth functions that vanish outside of Ω, we define the restricted fractional Laplacian $\left(-\Delta_{|\Omega}\right)^s$. For this operator, the most appropriate replacement for the Dirichlet boundary condition is $u \equiv 0$ in Ω^c.

The definitions given in (4) and (5), as well as several others detailed in Ref. [32], are equivalent to each other. One such definition, introduced in Ref. [33], involves the concept of harmonic extensions. This characterization of $(-\Delta)^s$ is described through the Dirichlet-to-Neumann map for a local degenerate elliptic PDE as follows. Let $u \in \mathcal{S}(\mathbb{R}^N)$. If $W = W(x, y)$, defined on $\mathbb{R}^N \times [0, \infty) \to \mathbb{R}$, is the unique solution to

$$\begin{cases} \operatorname{div}(y^{1-2s}\nabla W) = \Delta_x W + \dfrac{1-2s}{y} W_y + W_{yy} = 0 & \text{in } \mathbb{R}^N \times (0, \infty), \\ W(x, 0) = u(x) & \text{on } \mathbb{R}^N, \end{cases} \tag{6}$$

then,

$$(-\Delta)^s u(x) = \frac{4^s \Gamma(s)}{2\Gamma(1-s)} \lim_{y \to 0^+} \left(-y^{1-2s} W_y(x, y)\right). \tag{7}$$

This distinctive class of degenerate-singular problems, initially introduced in the 1980s by Fabes, Jerison, Kenig, and Serapioni [34–36], has gained significant attention in recent studies on fractional powers of the Laplacian.

For further details on the fractional Laplacian, refer to Ref. [37–39].

It is widely recognized that for problems involving L^1 or measure data, the notion of a (weak) distributional solution does not guarantee uniqueness for the following types of problems:

$$\begin{cases} -\operatorname{div}(a(x, u, \nabla u)) = f & \text{in } \Omega, \\ u = 0 & \text{on } \partial\Omega, \end{cases} \tag{8}$$

where f is a bounded Radon measure or a function in $L^1(\Omega)$, and $a : \Omega \times \mathbb{R} \times \mathbb{R}^N \to \mathbb{R}^N$ is a Carathéodory function, possibly nonlinear and satisfying some coerciveness and monotonicity assumptions. Efforts have been made to identify additional conditions on distributional solutions to ensure both existence and uniqueness, leading to some parallel developments, as discussed in [40]. Although Stampacchia's definition of solution, as outlined in Ref. [41], does ensure uniqueness, it necessitates stronger conditions on the solution. In particular, his definition uses a wider class of test functions rather than just $C_c^\infty(\Omega)$. In Ref. [42], the concept of entropy solutions was initially defined for L^1 data and subsequently extended to include particular types of measures, as elaborated in Ref. [43]. Dall'Aglio introduced the concept of SOLA solutions in Ref. [44], whereas Lions and Murat, in Refs. [45, 46], advanced the idea of renormalized solutions.

Given that the problem includes a measure term as a source, it is appropriate to utilize the concept of entropy solutions, which will be defined in detail later in Section 2.

This note is organized as follows: Section 2 outlines the functional setting and defines the concepts of weak and entropy solutions for the problem, followed by the presentation of the existence theorems. Section 3 provides the proofs of these existence results and includes remarks on possible relaxations of the assumptions. Lastly, we include a brief Conclusion in Section 4.

2. Setting of the problem and the existence theorems

For any $1 \le p < \infty$ and $0 < s < 1$, the Sobolev-Slobodeckij space is defined as follows:

$$W^{s,p}(\Omega):=\left\{u\in L^p(\Omega):\int_\Omega\int_\Omega\frac{|u(x)-u(y)|^p}{|x-y|^{N+ps}}\,dx\,dy<\infty\right\},\tag{9}$$

where N denotes the dimension of the domain Ω. This space is equipped with the classical Gagliardo norm:

$$\|u\|_{W^{s,p}(\Omega)}=\|u\|_{L^p(\Omega)}+\left(\int_\Omega\int_\Omega\frac{|u(x)-u(y)|^p}{|x-y|^{N+ps}}\,dx\,dy\right)^{\frac{1}{p}}.\tag{10}$$

Also, $W_0^{s,p}(\Omega)$ is defined as the closure of $C_c^\infty(\Omega)$ in $W^{s,p}(\Omega)$.

It is well established that for $0<s\leq\frac{1}{p}$, $W_0^{s,p}(\Omega)=W^{s,p}(\Omega)$, whereas for $s>\frac{1}{p}$, the inclusion $W_0^{s,p}(\Omega)\subset W^{s,p}(\Omega)$ is strict. For more details, see for example ([47], Theorem 1.4.2.4).

Given a bounded smooth domain Ω in $\mathbb{R}^N$, we define the space

$$X^{s,p}(\Omega)=\left\{u:\mathbb{R}^N\to\mathbb{R}\,|u|_\Omega\in L^p(\Omega),\iint_{D_\Omega}\frac{|u(x)-u(y)|^p}{|x-y|^{N+ps}}\,dx\,dy<\infty\right\},\tag{11}$$

where $D_\Omega=\mathbb{R}^N\times\mathbb{R}^N\backslash(\Omega^c\times\Omega^c)$. With the norm defined as

$$\|u\|_{X^{s,p}(\Omega)}=\left(\int_\Omega|u|^p\,dx+\iint_{D_\Omega}\frac{|u(x)-u(y)|^p}{|x-y|^{N+ps}}\,dx\,dy\right)^{\frac{1}{p}},\tag{12}$$

this space becomes a Banach space.

When $p=2$, we denote the space $X^{s,2}(\Omega)$ as $X^s(\Omega)$. In this case, $X^s(\Omega)$ becomes a Hilbert space with the inner product defined by:

$$\langle u,v\rangle_{X^s(\Omega)}=\int_\Omega uv\,dx+\iint_{D_\Omega}\frac{(u(x)-u(y))(v(x)-v(y))}{|x-y|^{N+2s}}\,dx\,dy.\tag{13}$$

We also define the space $X_0^{s,p}(\Omega)$ as:

$$X_0^{s,p}(\Omega):=\left\{u\in X^{s,p}(\Omega)|u=0\,\text{almost everywhere in}\,\Omega^c\right\}.\tag{14}$$

We also denote $X_0^s(\Omega)$ as $X_0^{s,2}(\Omega)$. It is straightforward to observe that:

$$\left(\iint_{D_\Omega}\frac{|u(x)-u(y)|^p}{|x-y|^{N+ps}}\,dx\,dy\right)^{\frac{1}{p}}=\left(\int_{\mathbb{R}^N}\int_{\mathbb{R}^N}\frac{|u(x)-u(y)|^p}{|x-y|^{N+ps}}\,dx\,dy\right)^{\frac{1}{p}},\quad\forall u\in X_0^{s,p}(\Omega).\tag{15}$$

This equality introduces an alternative norm, which is equivalent to the norm given in (12) for $X_0^{s,p}(\Omega)$. We denote this norm by $\|u\|_{X_0^{s,p}(\Omega)}$, defined as:

$$\|u\|_{X_0^{s,p}(\Omega)}=\left(\iint_{D_\Omega}\frac{|u(x)-u(y)|^p}{|x-y|^{N+ps}}\,dx\,dy\right)^{\frac{1}{p}}.\tag{16}$$

There exists a positive constant C such that the following inequalities are valid for all $u \in X_0^{s,p}(\Omega)$:

$$\|u\|_{X_0^{s,p}(\Omega)} \le \|u\|_{W^{s,p}(\mathbb{R}^N)} \le C\|u\|_{X_0^{s,p}(\Omega)}. \tag{17}$$

It is important to note that $X_0^{s,p}(\Omega)$ can also be characterized as the closure of $C_c^\infty(\Omega)$ in $X^{s,p}(\Omega)$. Additionally, in the case of the Hilbert space ($p = 2$), the following holds:

$$\|u\|^2_{X_0^s(\Omega)} = 2C_{N,s}^{-1}\|(-\Delta)^{\frac{s}{2}}u\|^2_{L^2(\mathbb{R}^N)}, \tag{18}$$

where $C_{N,s}$ is the constant previously defined in the context of the fractional Laplacian.

See Refs. [38] and ([48], subsection 2.2) for proofs of the above facts.

Finally, it is noteworthy that using the fractional Hardy-Sobolev inequality (see, e.g., [47], Theorem 1.4.4.3), one can verify that:

$$X_0^{s,p}(\Omega) = \begin{cases} W^{s,p}(\Omega) & \text{for } 0 < s < \dfrac{1}{p}, \\ W_0^{s,p}(\Omega) & \text{for } \dfrac{1}{p} < s < 1, \end{cases} \tag{19}$$

and in the case $s = \frac{1}{p}$,

$$X_0^{\frac{1}{p},p}(\Omega) := \left\{ u \in W^{\frac{1}{p},p}(\Omega) : \int_\Omega \frac{u^p(x)}{\delta^{2p}(x)}\,dx < +\infty \right\}, \tag{20}$$

where $\delta(x) := \text{dist}(x, \partial\Omega)$ is the distance function to the boundary. See, for example, ([47], Corollary 1.4.4.10).

We define $X_{\text{loc}}^{s,p}(\Omega)$ as the set of all functions u such that for every $\phi \in C_c^\infty(\Omega)$, the product $u\phi \in X_0^{s,p}(\Omega)$. When we say that a sequence $\{u_n\} \subset X_{\text{loc}}^{s,p}(\Omega)$ is (uniformly) bounded, we mean that for any fixed $\phi \in C_c^\infty(\Omega)$, the sequence $\{\phi u_n\} \subset X_0^{s,p}(\Omega)$ is bounded.

The Marcinkiewicz space $M^r(\Omega)$, for any fixed $0 < r < \infty$, is defined as the set of all measurable functions $w : \Omega \to \mathbb{R}$ for which there exists a constant $C > 0$ such that:

$$\omega(\{x \in \Omega : |w(x)| \ge t\}) \le \frac{C}{t^r}, \qquad \forall t > 0, \tag{21}$$

where ω denotes the Lebesgue measure on $\mathbb{R}^N$. This space is equipped with the seminorm:

$$\|w\|_{M^r(\Omega)} = \sup_{t>0} t(\omega(\{x \in \Omega : |w(x)| \ge t\}))^{\frac{1}{r}}. \tag{22}$$

We now recall the following continuous embedding (see, e.g., [42]):

$$L^r(\Omega) \rightarrowtail M^r(\Omega) \rightarrowtail L^{r-\varepsilon}(\Omega), \tag{23}$$

for any $1 < r < \infty$ and any $0 < \varepsilon \le r - 1$. Moreover, we will utilize the following continuous embedding:

$$X_0^{s,p}(\Omega) \rightarrowtail L^t(\Omega), \qquad \forall t \in [1, p_s^*]. \tag{24}$$

Here, $p_s^* = \frac{pN}{N-ps}$ denotes the Sobolev critical exponent, and as usual, this embedding is compact for any $t < p_s^*$. See, e.g., ([38], Theorem 6.5 and Theorem 7.1).

Given that we are working with the non-local operator $(-\Delta)^s$, it is necessary to introduce a new class of test functions more rigorously than the usual Schwartz test functions $C_c^\infty(\Omega)$. Specifically, we need to consider:

$$\mathcal{T} = \{\phi : \mathbb{R}^N \to \mathbb{R} \mid (-\Delta)^s \phi = \varphi, \varphi \in C_c^\infty(\Omega), \phi = 0 \text{ in } \mathbb{R}^N \backslash \tilde{\Omega}, \text{for some } \tilde{\Omega} \Subset \Omega\}. \tag{25}$$

It is known that $\mathcal{T} \subset X_0^s(\Omega) \cap L^\infty(\Omega)$; moreover, for any $\phi \in \mathcal{T}$, there exists a constant $\beta \in (0,1)$ such that $\phi \in C^{0,\beta}(\Omega)$. See Refs. [7, 42, 49].

For $u \in X_0^s(\Omega)$ and $\phi \in \mathcal{T}$, the following holds:

$$\begin{aligned} 2C_{N,s}^{-1} \int_{\mathbb{R}^N} u(-\Delta)^s \phi dx &= 2C_{N,s}^{-1} \int_{\mathbb{R}^N} (-\Delta)^{\frac{s}{2}} u (-\Delta)^{\frac{s}{2}} \phi dx \\ &= \iint_{D_\Omega} \frac{(u(x) - u(y))(\phi(x) - \phi(y))}{|x-y|^{N+2s}} dx dy. \end{aligned} \tag{26}$$

This equation establishes the self-adjointness of $(-\Delta)^s$ in $X_0^s(\Omega)$. Additionally, it can be shown that $(-\Delta)^s : X_0^s(\Omega) \to X^{-s}(\Omega)$ is a continuous and strictly monotone operator, where $X^{-s}(\Omega)$ is the dual space of $X_0^s(\Omega)$.

Definition 2.1 The function $u \in L^1_{\text{loc}}(\Omega)$ is a weak solution to (1) if:

- $u = 0$ in Ω^c;
- for any compact set K in Ω, there exists $C_K > 0$ such that $u(x) \geq C_K$ almost everywhere in K;
- Eq. (1) is fulfilled in the sense of distributions using the test functions in $\mathcal{T}$, i.e.,

$$\int_{\mathbb{R}^N} u(-\Delta)^s \phi dx = \int_\Omega \frac{\mu}{u^\gamma} \phi dx + \int_\Omega \phi d\nu, \qquad \forall \phi \in \mathcal{T}. \tag{27}$$

Since ϕ has compact support in Ω, the first term on the right-hand side of (27) is well-defined for any $\mu \in L^1_{\text{loc}}(\Omega)$, because u is positive on compact subsets of Ω. Here, $d\nu$ represents the weighted Lebesgue measure νdx for $\nu \in L^1_{\text{loc}}(\Omega)$.

Regarding uniqueness, we introduce an alternative definition of solutions to (1). Specifically, we aim to define the concept of an entropy solution. The inspiration for this definition is drawn from Ref. [43]. We denote the standard truncation operator by

$$T_k(s) = \begin{cases} s & \text{if } |s| \leq k \\ k \operatorname{sign}(s) & \text{if } |s| \geq k. \end{cases} \tag{28}$$

Definition 2.2 For a measure ν belonging to $L^1(\Omega) + X^{-s}(\Omega)$, the function $u \in L^1(\Omega)$ is an entropy solution to (1) if:

- $u = 0$ in Ω^c;
- for any compact set K in Ω, there exists $C_K > 0$ such that $u(x) \geq C_K$ almost everywhere in K;
- for every k, $T_k(u)$ belongs to $X_0^s(\Omega)$, and u meets the following set of inequalities:

$$\int_{\{|u-\phi|<k\}} (-\Delta)^{\frac{s}{2}} u (-\Delta)^{\frac{s}{2}} (u-\phi)\,dx \leq \int_\Omega \frac{\mu}{u^\gamma} T_k(u-\phi)\,dx + \int_\Omega T_k(u-\phi)\,d\nu, \qquad (29)$$

for every k and for any ϕ in $X_0^s(\Omega) \cap L^\infty(\Omega)$.

Remark 1.1 Notice that $\int_\Omega T_k(u-\phi)\,d\nu$ is the simple notation for the duality pairing between $X_0^s(\Omega) \cap L^\infty(\Omega)$ and $L^1(\Omega) + X^{-s}(\Omega)$; that is,

$$\int_\Omega T_k(u-\phi)\,d\nu = \langle T_k(u-\phi), \nu \rangle_{X_0^s(\Omega)\cap L^\infty(\Omega), L^1(\Omega)+X^{-s}(\Omega)}. \qquad (30)$$

We will later demonstrate that the first term on the right-hand side is well-defined for certain specific data μ. Additionally, the assumption that $\nu \in L^1(\Omega) + X^{-s}(\Omega)$ guarantees the well-posedness of the last term on the right-hand side, since $T_k(u-\phi) \in X_0^s(\Omega) \cap L^\infty(\Omega)$. It is crucial to note that $L^1(\Omega) + X^{-s}(\Omega)$ is embedded in the dual space of $X_0^s(\Omega) \cap L^\infty(\Omega)$.

Theorem 1.2 For any $\gamma > 0$, $s \in (0,1)$, and any data $0 \leq \mu, \nu \in L^1(\Omega)$, there exists a positive weak solution to problem (1). More precisely:

i. When $0 < \gamma \leq 1$, then u belongs to $X_0^{s_1,p}(\Omega)$, for every $s_1 < s$, and for all $p < \frac{N}{N-s}$. Additionally, for any $k > 0$, $T_k(u) \in X_0^s(\Omega)$.

ii. When $\gamma > 1$, then u belongs to $X_{\text{loc}}^{s_1,p}(\Omega)$, for all $s_1 < s$, and for all $p < \frac{N}{N-s}$. Furthermore, $T_k^{\frac{\gamma+1}{2}}(u) \in X_0^s(\Omega)$, and $T_k(u) \in X_{\text{loc}}^s(\Omega)$, for any $k > 0$.

Besides, for the case of $0 < \gamma \leq 1$, under the following extra hypothesis:

$$\int_\Omega \mu \delta^{-s\gamma}\,dx < +\infty, \qquad (31)$$

where δ represents $\text{dist}(x, \partial\Omega)$, then this weak solution is also the unique positive entropy solution to the problem (1).

Theorem 1.3 Under the conditions of Theorem 1.2, if we additionally assume that

$$\int_\Omega \mu^2 \delta^{2s(1-\gamma)}\,dx < +\infty, \qquad (32)$$

then the uniqueness of the weak solution is ensured. More specifically:

i. For $\gamma > 1$, the weak solution derived in Theorem 1.2 is unique.

ii. For $0 < \gamma \leq 1$, the weak and the entropy solutions obtained in Theorem 1.2 are equivalent.

3. Proof of theorem 2 and theorem 3

Initially, we examine the following straightforward problem:

$$\begin{cases} (-\Delta)^s v = g & \text{in}\,\Omega, \\ v > 0 & \text{in}\,\Omega, \\ v = 0 & \text{in}\,\Omega^c, \end{cases} \tag{33}$$

where $0 \lneqq g \in L^\infty(\Omega)$. The existence and uniqueness of a (weak) energy solution to problem (33) is well-established. For the convenience of the reader, we summarize this result in the following proposition. Before stating the proposition, we need to define the set $\mathcal{C}$, which contains all functions $v \in L^\infty(\Omega)$ for which there exist constants $k_1, k_2 > 0$ such that

$$k_1 \delta(x)^s \le v(x) \le k_2 \delta(x)^s. \tag{34}$$

For simplicity in notation, without emphasizing the constants, we write $v \sim \delta^s$, particularly near the boundary.

Proposition 1.4 (For example, see Ref. [50]) If $g \in L^\infty(\Omega)$, $g \gneqq 0$, and $s \in (0,1)$, then the existence of a unique energy solution to problem (33) is guaranteed within the space $X_0^s(\Omega) \cap \mathcal{C} \cap C^{0,s}(\mathbb{R}^N)$.

The definition of a solution to (33) is as follows: A function $u \in X_0^s(\Omega)$ is regarded as a (weak) energy solution to problem (33) if

- $u = 0$ in Ω^c;
- the following holds:

$$\int_{\mathbb{R}^N} (-\Delta)^{\frac{s}{2}} u (-\Delta)^{\frac{s}{2}} \phi dx = \int_\Omega g \phi dx, \tag{35}$$

for every $\phi \in X_0^s(\Omega)$.

Now, we continue as follows. For any $v \in L^2(\Omega)$, define $\Phi(v) = w$, where w is the (energy) solution to the following problem for a fixed n:

$$\begin{cases} (-\Delta)^s w = \mu_n \left(|v| + \frac{1}{n}\right)^{-\gamma} + \nu_n & \text{in}\,\Omega, \\ w > 0 & \text{in}\,\Omega, \\ w = 0 & \text{in}\,\Omega^c. \end{cases} \tag{36}$$

Here, $\mu_n = T_n(\mu)$ and $\nu_n = T_n(\nu)$ denote the truncations at level n. If we can show that

$$\Phi : L^2(\Omega) \to L^2(\Omega) \tag{37}$$

has a fixed point $w_n \in L^2(\Omega)$, then this w_n will be the solution to the following problem in $X_0^s(\Omega) \cap \mathcal{C}$:

$$\begin{cases} (-\Delta)^s w_n = \dfrac{\mu_n}{\left(w_n + \frac{1}{n}\right)^\gamma} + \nu_n & \text{in}\,\Omega, \\ w_n > 0 & \text{in}\,\Omega, \\ w_n = 0 & \text{in}\,\Omega^c. \end{cases} \tag{38}$$

To accomplish this, we will use Schauder's fixed-point theorem (see, e.g., [51], Theorem 3.2.20). Specifically, we need to demonstrate the following:

1. *Continuity*: Show that Φ is continuous.

2. *Compactness*: Show that Φ is a compact operator.

3. *Invariant set*: Identify a bounded convex subset of $L^2(\Omega)$ that is invariant under Φ.

For continuity, consider a sequence $v_k \to v$ in $L^2(\Omega)$. It is straightforward to show that for each n:

$$\left\| \left(\mu_n \left(|v_k| + \frac{1}{n} \right)^{-\gamma} \right) - \left(\mu_n \left(|v| + \frac{1}{n} \right)^{-\gamma} \right) \right\|_{L^2(\Omega)} \to 0 \quad \text{as } k \to \infty. \tag{39}$$

Therefore, by leveraging the uniqueness of the solution to (33), we can conclude that $\Phi(v_k) \to \Phi(v)$.

To establish compactness, consider $v \in L^2(\Omega)$ and let w be the solution to (36). If $\lambda_1^s(\Omega)$ represents the smallest eigenvalue of $(-\Delta)^s$ in $X_0^s(\Omega)$ (refer to Ref. [52], Proposition 9), then we have:

$$\lambda_1^s(\Omega) \int_\Omega w^2 dx \le \int_{\mathbb{R}^N} \left| (-\Delta)^{\frac{s}{2}} w \right|^2 dx. \tag{40}$$

Using $\phi = w$ as a test function in (36), we obtain:

$$\int_{\mathbb{R}^N} \left| (-\Delta)^{\frac{s}{2}} w \right|^2 dx = \int_\Omega \frac{\mu_n w}{\left(|v| + \frac{1}{n}\right)^\gamma} dx + \int_\Omega w \nu_n dx. \tag{41}$$

For the first term on the right-hand side of (41), we have:

$$\int_\Omega \frac{\mu_n}{\left(|v| + \frac{1}{n}\right)^\gamma} w dx \le n^{\gamma+1} \int_\Omega w dx \le C_1 \left(\int_\Omega w^2 dx \right)^{\frac{1}{2}}, \tag{42}$$

where we used Hölder's inequality in the last step. Applying Hölder's inequality again, we obtain $\int_\Omega \nu_n w dx \le C_2 \left(\int_\Omega w^2 dx \right)^{\frac{1}{2}}$ for some constant $C_2 > 0$. Combining this inequality with (41) and (42), we obtain

$$\|w\|^2_{X_0^s(\Omega)} \le C_1 \|w\|_{X_0^s(\Omega)} + C_4 \left(\int_\Omega |w|^2 dx \right)^{\frac{1}{2}}. \tag{43}$$

This inequality, together with the embedding (24), implies that $\|w\|_{X_0^s(\Omega)} = \|\Phi(v)\|_{X_0^s(\Omega)}$ is uniformly bounded for $v \in L^2(\Omega)$. Thus, (40) implies that $\Phi(L^2(\Omega))$ is contained in a ball of finite radius in $L^2(\Omega)$. Therefore, this ball is invariant under Φ. Furthermore, by utilizing the compactness of the embedding (24) and the uniform boundedness of $\|\Phi(v)\|_{X_0^s(\Omega)}$, it follows that $\Phi(L^2(\Omega))$ is relatively compact in $L^2(\Omega)$. This concludes the proof that Φ is compact.

Proposition 1.5 Let $\{w_n\}$ denote the solutions to (38). The following a priori estimates hold:

i. If $0 < \gamma \leq 1$, then the sequence $\{T_k(w_n)\}_{n=1}^{\infty}$ is bounded in $X_0^s(\Omega)$.

ii. If $\gamma > 1$, then the sequence $\left\{T_k^{\frac{\gamma+1}{2}}(w_n)\right\}_{n=1}^{\infty}$ is bounded in $X_0^s(\Omega)$, and the sequence $\{T_k(w_n)\}_{n=1}^{\infty}$ is bounded in $X_{\text{loc}}^s(\Omega)$.

Proof: We follow the approach outlined in ([49], Theorem 5.2).
First case: boundedness of $\{T_k(w_n)\}_{n=1}^{\infty}$ in $X_0^s(\Omega)$ for $0 < \gamma \leq 1$.
By testing (38) with $\phi = T_k(w_n)$ and using ([42], Proposition 3), we obtain:

$$\int_{\mathbb{R}^N} \left|(-\Delta)^{\frac{s}{2}} T_k(w_n)\right|^2 dx \leq \int_\Omega \frac{\mu_n(x)}{\left(w_n + \frac{1}{n}\right)^\gamma} T_k(w_n)\, dx + \int_\Omega T_k(w_n)\nu_n\, dx. \tag{44}$$

For the first term on the right-hand side, we have:

$$\begin{aligned} \int_\Omega \frac{\mu_n(x)}{\left(w_n + \frac{1}{n}\right)^\gamma} T_k(w_n)\, dx &\leq \int_\Omega \frac{\mu_n(x) T_k^\gamma(w_n)}{\left(w_n + \frac{1}{n}\right)^\gamma} T_k^{1-\gamma}(w_n)\, dx \\ &\leq k^{1-\gamma} \|\mu_n\|_{L^1(\Omega)} = C_1, \end{aligned} \tag{45}$$

where we used the fact that $\frac{T_k^\gamma(w_n)}{\left(w_n+\frac{1}{n}\right)^\gamma} \leq \frac{w_n^\gamma}{\left(w_n+\frac{1}{n}\right)^\gamma} \leq 1$. For the last term, we have:

$$\int_\Omega T_k(w_n)\nu_n\, dx \leq k\|\nu_n\|_{L^1(\Omega)} = C_2. \tag{46}$$

Combining (44), (45), and (46), we obtain:

$$\int_{\mathbb{R}^N} \left|(-\Delta)^{\frac{s}{2}} T_k(w_n)\right|^2 dx \leq C_1 \|T_k(w_n)\|_{X_0^s(\Omega)} + C_3, \tag{47}$$

or equivalently,

$$\|T_k(w_n)\|_{X_0^s(\Omega)}^2 \leq C_1 \|T_k(w_n)\|_{X_0^s(\Omega)} + C_3, \tag{48}$$

which implies that $\{T_k(w_n)\}_{n=1}^{\infty}$ is bounded in $X_0^s(\Omega)$.
Second case: Boundedness of $\left\{T_k^{\frac{\gamma+1}{2}}(w_n)\right\}_{n=1}^{\infty}$ in $X_0^s(\Omega)$ and $\{T_k(w_n)\}_{n=1}^{\infty}$ in $X_{\text{loc}}^s(\Omega)$.
By using $\phi = T_k^\gamma(w_n)$ as a test function in (38), we derive:

$$\int_{\mathbb{R}^N} (-\Delta)^{\frac{s}{2}} w_n (-\Delta)^{\frac{s}{2}} T_k^\gamma(w_n)\, dx = \int_\Omega \frac{\mu_n(x)}{\left(w_n + \frac{1}{n}\right)^\gamma} T_k^\gamma(w_n)\, dx + \int_\Omega T_k^\gamma(w_n)\nu_n\, dx. \tag{49}$$

For the left-hand side, using (26) and the following elementary inequality:

$$(s_1 - s_2)\left(s_1^\gamma - s_2^\gamma\right) \geq \frac{4\gamma}{(\gamma+1)^2}\left(s_1^{\frac{\gamma+1}{2}} - s_2^{\frac{\gamma+1}{2}}\right)^2, \qquad \forall s_1, s_2 \geq 0, \gamma > 0, \tag{50}$$

we get:

$$\begin{aligned}
&\int_{\mathbb{R}^N}(-\Delta)^{\frac{s}{2}}w_n(-\Delta)^{\frac{s}{2}}T_k^{\gamma}(w_n)\,dx\\
&=\frac{C_{N,s}}{2}\iint_{D_\Omega}\frac{(w_n(x)-w_n(y))\left(T_k^{\gamma}(w_n)(x)-T_k^{\gamma}(w_n)(y)\right)}{|x-y|^{N+2s}}\,dx\,dy\\
&\geq\frac{2\gamma C_{N,s}}{(\gamma+1)^2}\iint_{D_\Omega}\frac{\left|T_k^{\frac{\gamma+1}{2}}w_n(x)-T_k^{\frac{\gamma+1}{2}}(w_n)(y)\right|^2}{|x-y|^{N+2s}}\,dx\,dy\\
&\geq C_4\int_{\mathbb{R}^N}\left|(-\Delta)^{\frac{s}{2}}T_k^{\frac{\gamma+1}{2}}(w_n)\right|^2dx.
\end{aligned}\tag{51}$$

We have:

$$\int_\Omega\frac{\mu_n(x)}{\left(w_n+\frac{1}{n}\right)^{\gamma}}T_k^{\gamma}(w_n)\,dx\leq\int_\Omega\frac{\mu_n(x)w_n^{\gamma}}{\left(w_n+\frac{1}{n}\right)^{\gamma}}\,dx\leq\|\mu_n\|_{L^1(\Omega)}=C_5,\tag{52}$$

$$\int_\Omega T_k^{\gamma}(w_n)\nu_n\,dx\leq k^{\gamma}\|\nu_n\|_{L^1(\Omega)}=C_6.\tag{53}$$

Combining (49), (51), (52), and (53), we obtain:

$$\int_{\mathbb{R}^N}\left|(-\Delta)^{\frac{s}{2}}T_k^{\frac{\gamma+1}{2}}(w_n)\right|^2dx\leq C_7\|T_k^{\frac{\gamma+1}{2}}(w_n)\|_{X_0^s(\Omega)}+C_8,\tag{54}$$

which implies the boundedness of $\left\{T_k^{\frac{\gamma+1}{2}}(w_n)\right\}_{n=1}^{\infty}$ in $X_0^s(\Omega)$.

Next, we show that $\{T_k(w_n)\}_{n=1}^{\infty}$ is bounded in $X_{\text{loc}}^s(\Omega)$.

To begin, observe that the strong maximum principle for the fractional Laplacian implies that for any compact set K in Ω, there exists $C(K)>0$ such that:

$$w_n(x)\geq C(K)>0,\qquad\text{almost everywhere in}\,K.\tag{55}$$

Therefore, we have

$$T_k(w_n)\geq\min\{k,C(K)\}=\tilde{C}.\tag{56}$$

Now, for any $(x,y)\in K\times K$, consider $\alpha_n:=\frac{T_k(w_n)(x)}{\tilde{C}}$ and $\beta_n:=\frac{T_k(w_n)(y)}{\tilde{C}}$. Since $\alpha_n,\beta_n\geq1$, we can use the following elementary inequality:

$$(\alpha_n-\beta_n)^2\leq\left(\alpha_n^{\frac{\gamma+1}{2}}-\beta_n^{\frac{\gamma+1}{2}}\right)^2.\tag{57}$$

Given the definitions of α_n and β_n, we obtain:

$$(T_k(w_n(x))-T_k(w_n(y)))^2\leq\tilde{C}^{1-\gamma}\left(T_k^{\frac{\gamma+1}{2}}(w_n(x))-T_k^{\frac{\gamma+1}{2}}(w_n(y))\right)^2.\tag{58}$$

Hence, the boundedness of $\{T_k(w_n)\}_{n=1}^{\infty}$ in $X_{\text{loc}}^s(\Omega)$ is a consequence of the boundedness of $\left\{T_k^{\frac{\gamma+1}{2}}(w_n)\right\}_{n=1}^{\infty}$ in $X_0^s(\Omega)$.

We now present the following proposition, which aligns with ([42], Theorem 23) and ([49], Theorem 4.10).

Proposition 1.6 (i) When $0 < \gamma \leq 1$, then $\{w_n\}_{n=1}^{\infty}$ is bounded uniformly in $X_0^{s_1,p}(\Omega)$ for any $s_1 < s$ and for any $p < \frac{N}{N-s}$.

(ii) When $\gamma > 1$, then $\{w_n\}_{n=1}^{\infty}$ is bounded uniformly in $X_{\text{loc}}^{s_1,p}(\Omega)$ for any $s_1 < s$ and any $p < \frac{N}{N-s}$.

Proof: The key point is that, for example in the first case of (i), by referring to the proofs of ([53], Proposition 3.4) and ([19], Lemma 2.8), it can be easily shown that:

- $\left\{(-\Delta)^{\frac{s}{2}} w_n\right\}_{n=1}^{\infty}$ is bounded in $M^{\frac{N}{N-s}}(\Omega)$ and then the embedding (23) implies the boundedness of $\left\{(-\Delta)^{\frac{s}{2}} w_n\right\}$ in $L^p(\Omega)$, for all $p < \frac{N}{N-s}$.

However, since $N > 2s$, we have $\frac{N}{N-s} < 2$. By applying Theorem 5 and Proposition 10 from Chapter 5 of Ref. [54], we conclude that $u \in X_0^{s_1,p}(\Omega)$ for all $s_1 < s$ and for all $p < \frac{N}{N-s}$. (In Ref. [54], $X_0^{s,p}(\Omega)$ corresponds to $\Lambda_s^{p,p}(\mathbb{R}^N)$, and $\mathcal{L}_s^p(\mathbb{R}^N)$ refers to the space of Bessel potentials; see ([54], Subsection 3.2).

We are now prepared to prove Theorem 1.2.

Proof of Theorem 1.2: By Proposition 1.6, we obtain the following results:

- **When** $0 < \gamma \leq 1$, then the sequence $\{w_n\}_{n=1}^{\infty}$ is uniformly bounded in $X_0^{s_1,p}(\Omega)$ for any $s_1 < s$ and any $p < \frac{N}{N-s}$. Hence, after possibly extracting a subsequence, w_n converges weakly to u in the space $X_0^{s_1,p}(\Omega)$. Also, $T_k(u) \in X_0^s(\Omega)$.

- **When** $\gamma > 1$, then the sequence $\{w_n\}_{n=1}^{\infty}$ is bounded in $X_{\text{loc}}^{s_1,p}(\Omega)$ for all $s_1 < s$ and for all $p < \frac{N}{N-s}$. Thus, after passing to a subsequence, w_n converges weakly to u in the local space $X_{\text{loc}}^{s_1,p}(\Omega)$. Moreover, $T_k^{\frac{\gamma+1}{2}}(u) \in X_0^s(\Omega)$ and $T_k(u) \in X_{\text{loc}}^s(\Omega)$.

Using results from embedding theorems, we can also assume, after possibly selecting a subsequence:

- If $0 < \gamma \leq 1$, then $w_n \to u$ in $L^r(\Omega)$ for any $r \in [1, 2_s^*)$.

- If $\gamma > 1$, then $w_n \to u$ in $L_{\text{loc}}^r(\Omega)$ for any $r \in [1, 2_s^*)$.

- The sequence $w_n(x)$ converges to $u(x)$ pointwise almost everywhere in Ω.

For any fixed $\phi \in \mathcal{T}(\Omega)$ (with compact support), we pass to the limit in the weak formulation to obtain:

$$\begin{aligned} \lim_{n\to\infty} \int_{\Omega} \frac{\mu_n}{\left(w_n + \frac{1}{n}\right)^{\gamma}} \phi dx &= \int_{\Omega} \frac{\mu}{u^{\gamma}} \phi dx, \\ \lim_{n\to\infty} \int_{\mathbb{R}^N} (-\Delta)^{\frac{s}{2}} w_n (-\Delta)^{\frac{s}{2}} \phi dx &= \int_{\mathbb{R}^N} u(-\Delta)^s \phi dx. \end{aligned} \tag{59}$$

Additionally, since $\nu_n \to \nu$ in $L^1(\Omega)$, we have:

$$\int_\Omega \nu_n \phi dx \to \int_\Omega \nu \phi dx. \tag{60}$$

Since, for any compact set K in Ω, there is a constant $C_K > 0$ such that $w_n(x) \geq w_1(x) \geq C_K$ almost everywhere in K, and since $w_n \equiv 0$ outside Ω, along with $w_n(x) \to u(x)$ almost everywhere in Ω, we can deduce that u is a weak solution to the problem (1). This finalizes the proof of the existence theorem.

As mentioned before, for uniqueness, we will consider the entropy solution.

Consider the following approximation problem:

$$\begin{cases} (-\Delta)^s u_n = \mu_n \left(u_n + \frac{1}{n}\right)^{-\gamma} + \nu_n & \text{in}\, \Omega, \\ u_n > 0 & \text{in } \Omega, \\ u_n = 0 & \text{in}\, \Omega^c. \end{cases} \tag{61}$$

Now, let us consider the sequence $\{T_k(u_n - \phi)\}_{n=1}^\infty$ for any $\phi \in X_0^s(\Omega) \cap L^\infty(\Omega)$ and for any fixed k. According to Proposition 5, this sequence is bounded in $X_0^s(\Omega)$. Consequently, after possibly taking a subsequence, $T_k(u_n - \phi)$ converges weakly to $T_k(u - \phi)$ in $X_0^s(\Omega)$, where u is the weak solution to (1).

Additionally, $\{T_k(u_n - \phi)\}_{n=1}^\infty$ is an increasing sequence of non-negative functions due to the strict monotonicity of the operator $(-\Delta)^s u$ and the non-decreasing nature of $\mu_n \left(u_n + \frac{1}{n}\right)^{-\gamma} + \nu_n$. The strict monotonicity of $(-\Delta)^s$ implies $T_k(u_n - \phi) \uparrow T_k(u - \phi)$, and consequently, $T_k(u_n - \phi) \to T_k(u - \phi)$ strongly in $X_0^s(\Omega)$ (as detailed in Ref. ([49], Lemmas 2.18).

Moreover, we have:

$$\left| \frac{\mu_n T_k(u_n - \phi)}{\left(u_n + \frac{1}{n}\right)^\gamma} \right| \leq k \frac{\mu}{u_1^\gamma} \in L^1(\Omega), \tag{62}$$

since $u_1 \sim \delta^s$, and as a result of (31).

Finally, by testing Eq. (61) with $T_k(u_n - \phi)$ and passing to the limit, we obtain an entropy solution that also satisfies the equalities in Definition 2.2.

To prove uniqueness, let u and v be two entropy solutions. We test u with $\phi = T_h(v)$ and v with $\phi = T_h(u)$ in the weak formulation of the entropy inequalities. By following the method outlined in Ref. ([55], Section 5) and utilizing the monotonicity of the nonlinear term $u^{-\gamma}$ with respect to u, the uniqueness of the solution follows almost immediately. See, for example, ([8], Section 4). This concludes the proof.

Remark 1.7 The proof of the existence of a positive weak solution can be generalized as follows:

- **Generalization for the bounded Radon measure** ν: Instead of defining ν_n as $T_n(\nu)$, we can consider ν_n as a sequence of non-negative functions in $L^\infty(\Omega)$ that are bounded in $L^1(\Omega)$ and converge to ν in the narrow topology of measures. This means that

$$\int_\Omega \phi \nu_n dx \to \int_\Omega \phi d\nu, \quad \forall \phi \in C_c(\Omega). \tag{63}$$

The same proof technique can be applied in this more generalized context.

- **Generalization for** μ: Instead of assuming $\mu \in L^1(\Omega)$, we can consider $\mu \in L^1(\Omega) + X^{-s}(\Omega)$, provided that ν is sufficiently regular to ensure that the solution u belongs to $X^s_{loc}(\Omega)$. In this scenario, μ_n should be a sequence of non-negative functions in $X^s_0(\Omega) \cap L^\infty(\Omega)$ that are bounded in $L^1(\Omega)$ and converge to μ in the sense that

$$\int_\Omega \phi \mu_n dx \to \langle \phi, \mu \rangle_{X^s_0(\Omega)\cap L^\infty(\Omega), L^1(\Omega)+X^{-s}(\Omega)}, \quad \forall \phi \in X^s_0(\Omega) \cap L^\infty(\Omega). \tag{64}$$

For further relaxations on the assumptions on μ, see ([56], Section 4).

Remark 8 For the existence of a positive entropy solution, it is essential that the approximating sequences μ_n and ν_n be increasing. This requirement can typically be met using the standard truncation technique for $\mu, \nu \in L^1(\Omega)$. However, it is important to note that non-negative bounded Radon measures cannot generally be approximated by increasing sequences of bounded functions, as discussed in Ref. [9].

For further details on approximating measures and exploring more relaxed assumptions on μ, see ([56], Section 4). This reference provides insights into approximations by increasing sequences of bounded functions for measures $\mu \in L^1(\Omega) + X^{-s}(\Omega)$.

Proof of Theorem 1.3:

Uniqueness for $\gamma > 1$.

Let u_1 and u_2 be two positive weak solutions to problem (1). Since $T_k^{\frac{\gamma+1}{2}}(u_i) \in X^s_0(\Omega)$ for $i = 1, 2$, define $w = u_1 - u_2$. Then we have

$$\int_{\mathbb{R}^N} w(-\Delta)^s \phi dx = \int_\Omega \left(\frac{\mu}{u_1^\gamma} - \frac{\mu}{u_2^\gamma} \right) \phi dx, \quad \forall \phi \in \mathcal{T}(\Omega). \tag{65}$$

Alternatively, note that by applying Hölder's inequality, the fractional Hardy-Sobolev inequality, and using (34), along with the comparison principle for the fractional Laplacian and a density argument, equality (65) holds for all $\phi \in X^s_0(\Omega) \cap L^\infty(\Omega)$, provided that assumption (32) is met.

Using $T_k^{\frac{\gamma+1}{2}}(w^-)$ as a test function in (65) and applying (26) and (50), we get

$$\int_{\mathbb{R}^N} \left| (-\Delta)^{\frac{s}{2}} T_k^{\frac{\gamma+3}{4}}(w^-) \right|^2 dx \le \frac{(\gamma+3)^2}{4(\gamma+1)C_{N,s}} \left\{ \int_\Omega \left(\frac{\mu}{u_1^\gamma} - \frac{\mu}{u_2^\gamma} \right) T_k^{\frac{\gamma+1}{2}}(w^-) dx \right\} \le 0. \tag{66}$$

Thus, $T_k^{\frac{\gamma+3}{4}}(w^-) \equiv 0$. This implies that $T_k^{\frac{\gamma+3}{4}}(u_1) \ge T_k^{\frac{\gamma+3}{4}}(u_2)$. Similarly, $T_k^{\frac{\gamma+3}{4}}(u_1) \le T_k^{\frac{\gamma+3}{4}}(u_2)$, which leads to $T_k^{\frac{\gamma+3}{4}}(u_1) = T_k^{\frac{\gamma+3}{4}}(u_2)$ for all k. Hence, uniqueness follows.

Uniqueness for $0 < \gamma \le 1$.

Let u_1 and u_2 be two positive weak solutions to problem (1) and define $w = u_1 - u_2$. Since $T_k(w) \in X^s_0(\Omega)$, using $T_k(w^-)$ as a test function in (65), we obtain

$$\int_{\mathbb{R}^N} \left| (-\Delta)^{\frac{s}{2}} T_k(w^-) \right|^2 dx \le \int_\Omega \left(\frac{\mu}{u_1^\gamma} - \frac{\mu}{u_2^\gamma} \right) T_k(w^-) dx \le 0. \tag{67}$$

Therefore, $T_k(w^-) \equiv 0$. Similarly, $T_k(w^+) \equiv 0$ for any k, which implies that $w \equiv 0$ almost everywhere in Ω. Thus, $u_1 = u_2$ in Ω, and hence, uniqueness follows. Consequently, the notions of the entropy solution and the weak solution coincide in this case.

4. Conclusion

In this chapter, we have explored the existence and uniqueness of positive solutions to a singular fractional elliptic problem governed by the fractional Laplacian, involving singular nonlinearities and irregular data. By employing both weak and entropy solutions, we demonstrated the existence of solutions for a wide range of cases, including the important scenario of measure data. For $0 < \gamma \leq 1$, we established the existence and uniqueness of entropy solutions, proving their equivalence to weak solutions in this range. We also showed uniqueness for weak solutions when $\gamma > 1$, extending the understanding of fractional Laplacian problems with singular behavior.

The mathematical framework we employed, involving the restricted fractional Laplacian and the concept of entropy solutions, provided a robust structure for addressing the challenges posed by irregular data and singular nonlinearities. Moreover, the relaxation of certain assumptions on the data opens the door for broader applications of the results.

Future research could further investigate the relaxation of the regularity conditions on both the domain and the data, as well as explore other types of singular and degenerate problems within the framework of fractional operators. The interplay between local and non-local operators in such problems, particularly in applied contexts, also remains a compelling area for continued study.

Acknowledgements

M. Bayrami was supported by a grant from IPM.

Author details

Masoud Bayrami[1*†] and Mahmoud Hesaaraki[2†]

1 School of Mathematics, Institute for Research in Fundamental Sciences (IPM), Tehran, Iran

2 Department of Mathematical Sciences, Sharif University of Technology, Tehran, Iran

*Address all correspondence to: aminlouee@ipm.ir

† These authors contributed equally.

IntechOpen

© 2024 The Author(s). Licensee IntechOpen. This chapter is distributed under the terms of the Creative Commons Attribution License (http://creativecommons.org/licenses/by/4.0), which permits unrestricted use, distribution, and reproduction in any medium, provided the original work is properly cited.

References

[1] Lazer AC, McKenna PJ. On a singular nonlinear elliptic boundary-value problem. Proceedings of the American Mathematical Society. 1991;**111**(3):721-730

[2] Bougherara B, Giacomoni J, Hernández J. Existence and regularity of weak solutions for singular elliptic problems, 2014 Madrid Conference on Applied Mathematics in honor of Alfonso casal. Electronic Journal of Differential Equations. 2015;**22**:19-30

[3] Abdellaoui B, Attar A. Quasilinear elliptic problem with hardy potential and singular term. Communications on Pure and Applied Analysis. 2013;**12**(3):1363-1380

[4] Abdellaoui B, Fernández AJ. Nonlinear fractional laplacian problems with nonlocal 'gradient terms'. Proceedings of the Royal Society of Edinburgh Section A: Mathematics. 2020;**150**(5):2682-2718

[5] Alves CO, Santos CA, Siqueira TW. Uniqueness in Wloc1,pxΩ and continuity up to portions of the boundary of positive solutions for a strongly-singular elliptic problem. Journal of Differential Equations. 2020;**269**(12):11279-11327

[6] Arora R, Giacomoni J, Goel D, Sreenadh K. Symmetry of solutions to singular fractional elliptic equations and applications. Comptes Rendus. Mathématique. 2020;**358**(2):237-243

[7] Barrios B, De Bonis I, Medina M, Peral I. Semilinear problems for the fractional laplacian with a singular nonlinearity. Open Mathematics. 2015;**13**(1):390-407

[8] Bayrami-Aminlouee M, Hesaaraki M, Hamdani MK, Chung NT. Nonlocal Lazer–McKenna-type problem perturbed by the Hardy's potential and its parabolic equivalence. Boundary Value Problems. 2021;**2021**:1-42

[9] Boccardo L, Orsina L. Semilinear elliptic equations with singular nonlinearities. Calculus of Variations and Partial Differential Equations. 2010;**37**(3–4):363-380

[10] Canino A, Montoro L, Sciunzi B, Squassina M. Nonlocal problems with singular nonlinearity. Bulletin des Sciences Mathematiques. 2017;**141**(3):223-250

[11] Chipot M. On some singular nonlinear problems for monotone elliptic operators. Atti Accad. Naz. Lincei Rend. Lincei Mat. Appl. 2019;**30**(2):295-316

[12] Chipot M, De Cave LM. New techniques for solving some class of singular elliptic equations. Atti Accad. Naz. Lincei Rend. Lincei Mat. Appl. 2018;**29**(3):487-510

[13] Crandall MG, Rabinowitz PH, Tartar L. On a Dirichlet problem with a singular nonlinearity. Communications in Partial Differential Equations. 1977;**2**(2):193-222

[14] De Cave LM, Durastanti R, Oliva F. Existence and uniqueness results for possibly singular nonlinear elliptic equations with measure data. NoDEA. 2018;**25**:18

[15] De Cave LM, Oliva F. On the regularizing effect of some absorption and singular lower order terms in classical Dirichlet problems with L1 data. Journal of Elliptic and Parabolic Equations. 2016;**2**(1–2):73-85

[16] Durastanti R. Asymptotic behavior and existence of solutions for singular

elliptic equations. Annali di Matematica Pura ed Applicata (1923-). 2020;**199**: 925-954

[17] Esposito F, Sciunzi B. On the Höpf boundary lemma for quasilinear problems involving singular nonlinearities and applications. Journal of Functional Analysis. 2020;**278**(4): 108346

[18] Garain P, Mukherjee T. Quasilinear nonlocal elliptic problems with variable singular exponent. Communications on Pure and Applied Analysis. 2020;**19**(11): 5059-5075

[19] Ghosh S, Choudhuri D, Giri RK. Singular nonlocal problem involving measure data. Bulletin of the Brazilian Mathematical Society, (New Series). 2019;**50**(1):187-209

[20] Giacomoni J, Mukherjee T, Sreenadh K. Existence of three positive solutions for a nonlocal singular Dirichlet boundary problem. Advanced Nonlinear Studies. 2019;**19**(2):333-352

[21] Giacomoni J, Santra S, et al. Positive solutions to a fractional equation with singular nonlinearity. Journal of Differential Equations. 2018;**265**(4): 1191-1226

[22] Marta L, Oliva F, Petitta F, de León SS. The Dirichlet problem for the 1-laplacian with a general singular term and L1-data. Nonlinearity. 2021;**34**(3): 1791-1816

[23] Molica Bisci G, Repovš D. Existence and localization of solutions for nonlocal fractional equations. Asymptotic Analysis. 2014;**90**(3–4):367-378

[24] Oliva F. Regularizing effect of absorption terms in singular problems. Journal of Mathematical Analysis and Applications. 2019;**472**(1):1136-1166

[25] Oliva F, Petitta F. A nonlinear parabolic problem with singular terms and nonregular data. Nonlinear Analysis. 2020;**194**:111472

[26] Oliva F, Petitta F. On singular elliptic equations with measure sources. ESAIM. 2016;**22**(1):289-308

[27] Oliva F, Petitta F. Finite and infinite energy solutions of singular elliptic problems: Existence and uniqueness. Journal of Differential Equations. 2018; **264**(1):311-340

[28] Orsina L, Petitta F. A Lazer-McKenna type problem with measures. Differential and Integral Equations. 2016;**29**(1–2):19-36

[29] Panda A, Ghosh S, Choudhuri D. Elliptic partial differential equation involving a singularity and a radon measure. The Journal of the Indian Mathematical Society. 2019;**86**(1–2): 95-117

[30] Nachman A, Callegari AJ. A nonlinear singular boundary value problem in the theory of pseudoplastic fluids. SIAM Journal on Applied Mathematics. 1980;**38**(2):275-281

[31] Ferone A, Mercaldo A, de Le'on SS. Singular elliptic equations having a gradient term with natural growth. arXiv preprint arXiv:2401.06237. 2024

[32] Kwaśnicki M. Ten Equivalent definitions of the fractional Laplace operator. Fractional Calculus and Applied Analysis. 2017;**20**(1):7-51

[33] Caffarelli L, Silvestre L. An extension problem related to the fractional laplacian. Communications in Partial Differential Equations. 2007;**32** (8):1245-1260

[34] Fabes EB, Jerison D, Kenig C. Boundary behavior of solutions to

degenerate elliptic equations. In: Conference on Harmonic Analysis in Honor of Antoni Zygmund. Vol. 1. 1983. pp. 577-589

[35] Fabes EB, Jerison DS, Kenig CE. The wiener test for degenerate elliptic equations. In Annales de l'institut Fourier. 1982;**32**:151-182

[36] Fabes EB, Kenig CE, Serapioni RP. The local regularity of solutions of degenerate elliptic equations. Communications in Statistics-Theory and Methods. 1982;**7**(1):77-116

[37] Abatangelo N, Valdinoci E. Getting acquainted with the fractional Laplacian. In: Contemporary Research in Elliptic PDEs and Related Topics, Vol. 33, Springer INdAM Ser. Cham: Springer; 2019. pp. 1-105

[38] Di Nezza E, Palatucci G, Valdinoci E. Hitchhiker's guide to the fractional Sobolev spaces. Bulletin des Sciences Mathematiques. 2012;**136**(5):521-573

[39] Silvestre L. Regularity of the obstacle problem for a fractional power of the Laplace operator. Communications on Pure and Applied Mathematics. 2007;**60**(1):67-112

[40] Dal Maso G, Murat F, Orsina L, Prignet A. Renormalized solutions of elliptic equations with general measure data. Annali della Scuola Normale Superiore di Pisa-Classe di Scienze. 1999; **28**(4):741-808

[41] Stampacchia G. Le problème de dirichlet pour les équations elliptiques du second ordre à coefficients discontinus. Annales de l'institut Fourier 1965, vol. 15, pp. 189-257

[42] Leonori T, Peral I, Primo A, Soria F. Basic estimates for solutions of a class of nonlocal elliptic and parabolic equations. Discrete and Continuous Dynamical Systems. 2015;**35**(12):6031-6068

[43] Boccardo L, Gallouët T, Orsina L. Existence and uniqueness of entropy solutions for nonlinear elliptic equations with measure data. In: Annales de l'Institut Henri Poincaré C, Analyse non linéaire. Vol. 13. no. 5. Elsevier; 1996. pp. 539-551

[44] Dall'Aglio, A. Approximated solutions of equations with l 1 data. Application to the h-convergence of quasi-linear parabolic equations. Annali di Matematica Pura ed Applicata. 1996; **170**:207-240

[45] Lions P-L, Murat F. Sur les solutions renormalisées d'équations elliptiques non linéaires to appear; 1965. unpublished paper

[46] Murat F. Soluciones renormalizadas de EDP elipticas no lineales. preprint. 93023. 1993

[47] Grisvard P. Elliptic Problems in Nonsmooth Domains, Vol. 24 of Monographs and Studies in Mathematics. Boston, MA: Pitman (Advanced Publishing Program); 1985

[48] Servadei R, Valdinoci E. Mountain pass solutions for non-local elliptic operators. Journal of Mathematical Analysis and Applications. 2012;**389**(2): 887-898

[49] Abdellaoui B, Medina M, Peral I, Primo A. The effect of the hardy potential in some calderón–zygmund properties for the fractional laplacian. Journal of Differential Equations. 2016; **260**(11):8160-8206

[50] Ros-Oton X, Serra J. The dirichlet problem for the fractional laplacian: Regularity up to the boundary.

Journal de Mathématiques Pures et Appliquées. 2014;**101**(3):275-302

[51] Papageorgiou NS, Rădulescu VD, Repovš DD. Nonlinear analysis—Theory and methods. In: Springer Monographs in Mathematics. Cham: Springer; 2019

[52] Servadei R, Valdinoci E. Variational methods for non-local operators of elliptic type. Discrete and Continuous Dynamical Systems. 2013;**33**(5): 2105-2137

[53] Bayrami-Aminlouee M, Hesaaraki M. Existence of a unique positive entropy solution to a singular fractional laplacian. Complex Variables and Elliptic Equations. 2021;**66**(5):783-800

[54] Stein EM. Singular integrals and differentiability properties of functions. In: Princeton Mathematical Series, No. 30. Princeton, N.J: Princeton University Press; 1970

[55] Bénilan P, Boccardo L, Gallouët T, Gariepy R, Pierre M, Vázquez JL. An l1-theory of existence and uniqueness of solutions of nonlinear elliptic equations. Annals of the Scuola Normale Superiore of Pisa-Science Class. 1995;**22**(2):241-273

[56] Bayrami-Aminlouee M, Hesaaraki M. A fractional laplacian problem with mixed singular nonlinearities and nonregular data. Journal of Elliptic and Parabolic Equations. 2021;**7**:787-814

Chapter 5

Application of Fractional Integral Transform in Fuzzy Differential Equations

Noreen Azhar, Saleem Iqbal, Raheela Manzoor and Shazia Kalsoom

Abstract

This chapter has included the application of fractional integral transform in the fuzzy field by presenting the solution of fuzzy fractional differential equations with the help of different integral transforms Laplace, Mellin, and Sumudu fuzzy transforms. These integral transforms are properly defined with their operational properties and are particularly described in differential and integral fractional calculus. The main results and discussion of the present research study about the applications of fractional integral transform are analyzed in this chapter with analytical solutions. The key features of this work are to discuss different integral transforms in a fuzzy theory and name it fuzzy integral transform with all its basic concepts. A procedure is established to solve fuzzy initial value problems (FIVP) by the fractional Mellin transforms and fractional Laplace and fractional Sumudu integral transform techniques, by applying the differential properties of fractional integral transforms, under a mathematical framework to solve the several types of differential equations.

Keywords: fuzzy-valued function, fractional calculus, fuzzy integral transform, Riemann-Liouville fractional derivative and integral operator, fuzzy fractional differential equations

1. Introduction

Fractional calculus has grown in importance within the area of mathematical modeling, as non-integer-order derivatives and integrals are involved. Fractional calculus (FC) has been applied over the last two decades, especially in engineering, to represent various physical and chemical process problems in science [1–5]. Due to the fractional order operators, it is distinct from classical calculus. Fractional calculus extends the integer-order calculus that investigates integrals and derivatives of arbitrary real or complex order. It can be applied in various fields of engineering and sciences, such as rheology, fluid flow, viscoelasticity, electrical networks, and electrochemistry of corrosion [6–9]. Many researchers applied FC theoretically and practically, to solve fractional order differential equations that are ordinary differential equations expanded to any non-integer order. Agarwal et al. [1] provided the notion and introduction of fractional order differential equations. Kilbas and Lakshmikanthan examined fractional order differential equations from a basic theory perspective [8, 10]. Fractional order differential equations were also solved using a variety of integral transforms, primarily the Fourier and Laplace transforms. Their contribution to fractional calculus is illustrated by the use of the Riesz-Feller fractional derivatives, which are determined by the Fourier and inverse Fourier transforms [11]. Moreover, for the solution of fractional order differential equations, the Caputo and Riemann-Liouville fractional derivatives are defined by Laplace transform [12, 13]. Through the rationalities of fractional calculus, for the solution of fractional order differential equations, Riemann-Liouville and Caputo fractional derivatives are more prevalent and common. The significance and scope of the integral transform's application in Fourier analysis may be seen in the fact that Mellin convolution-type integral transforms can be used to represent the FC operators used in the Laplace integral transform as well as the sine and cosine Fourier transforms. Moreover, the scale preservation quality of the fractional Sumudu transforms makes the converted function identical to the original function [14–16].

In this chapter, a few practical applications of the integral transform in fuzzy fractional order differential equations (FDE) will be covered. Fuzzy differential equations help us to understand the behavior of uncertain dynamical systems, which enables us to design an adaptable control system that may be realistic and accurate.

Hukuhara derivatives of a fuzzy number-valued function are one method used by many researchers to solve FDE [17, 18]. The fuzzy differential equations were solved by Buckley and Feuring [19] using Zadeh's extension principle and strongly generalized differentiability notions in order to ascertain the existence of periodic solutions or asymptotic phenomena. Although both types of fuzzy differential equations lack exact solutions, numerical techniques are used to determine an approximate solution with error analysis. The fuzzy differential equation of fractional order can be investigated by assuming the same fundamental notions as classical fuzzy differential equations [20]. However, when several parameters are implicated in the mathematical modeling of natural processes, the result is an imprecise fractional model. A fractional model of an energy supply-demand system was recently solved by Noeiaghdam and Sidorov [21] using the Caputo-Fabrizio derivative. Fractional integral transform techniques, such as the Fourier and Laplace fractional integral transforms, can be used to solve differential equations of fractional order [22, 23]. In this chapter, we are able to solve fractional order fuzzy differential equations by using Fuzzy Laplace, Fuzzy Mellin, and Fuzzy Sumudu transform methods. Some of the limitations and drawbacks of fractional order differential equations are featured when they deal with the initial value that is assigned to the model. Generally, it involves uncertain quantities, and determining the initial values is absurdly difficult when it pacts with physical phenomena. Many scientists have discussed various novel ideas to address uncertainty quantities, and fuzzy set theory is one of the most promising of them [24]. Since the theory included the capacity to manage differential equations with initial value uncertainty, the field of fractional order differential equations with uncertainty was originally approached by Agarwal et al. [1]. Subsequently, many scientists were persuaded to investigate and broaden this topic even more [25–28]. For this, first we discuss some preliminaries of fuzzy sets fuzzy operator and fractional calculus on fuzzy numbers. The second section includes the definition of fuzzy fractional integral transform. Section 3 includes a definition of fuzzy fractional differential equation along with the procedure for determining the solution through fuzzy integral transform methods. In Section 4, some examples are elaborated for solving fuzzy differential equations of fractional by using different fuzzy integral transform methods.

1.1 Preliminaries

A fuzzy set is created by using ordered pairings where the second component is the first component's degree of membership. Here, fuzzy set is an uncertain set, T the membership of its elements, $x \in T$, has a membership degree, $T(x) \in [0, 1]$. If the degree is zero, the member does not belong to the set and if the degree is one, the member belongs to the set completely [29].

A fuzzy membership function $T(x) \in [0, 1]$, is called a fuzzy number if it has the following conditions:

i. T is normal, such that $T(x_0) = 1$ for $x_0 \in \mathbb{R}$

ii. T, is fuzzy convex. For any two arbitrary real points $x_1; x_2$ and $\mu \in [0, 1]$ we have

$$T(\mu x_1 + (1-\mu)\, x_2) \geq min\, [T(x_1), T(x_2)]$$

iii. T, is upper semi-continuous.

iv. The closure of the set is compact, i.e.,

$$Supp(T) = x \in \mathbb{R} | T(x) > 0$$

Fuzzy operations can be applied to any two arbitrary fuzzy integers, M and N, in different forms, such as parametric, trapezoidal, and triangular, and whenever considering a continuous measurable function such as $f \in \{\bigoplus, \bigodot, \varnothing\}$, the fuzziness of M and N is neither more than nor equal to the fuzziness of $f\ (M, N)$. Furthermore, the fuzziness of M at any level of α can be defined as the diameter of an interval in a fuzzy number. If f is continuous and measurable then for $0 < \alpha < 1$

$$\text{if } \mathrm{M} \oplus \mathrm{N} = \mathrm{P} \text{ then } \mathrm{M}(\alpha) + \mathrm{N}(\alpha) = \mathrm{P}(\alpha), \text{(addition)}$$
$$\text{if } \lambda \bigodot \mathrm{n} = \mathrm{P} \text{ then } \lambda \bigodot \mathrm{N}(\alpha) = \mathrm{P}(\alpha), \text{(Scalar multiplication)}$$
$$\text{if } \mathrm{M} \odot \mathrm{N} = \mathrm{P} \text{ then } \mathrm{M}(\alpha) \odot \mathrm{N}(\alpha) = \mathrm{P}(\alpha), \text{(multiplication)}$$

A fuzzy-valued function is an extension of a classical function that takes ambiguity and uncertainty into account. It is represented by a membership function, usually written as u(x), which gives each element in the domain a degree of membership in the interval [0,1].

Moreover, H-differentiability for fuzzy-valued functions is further defined by defining Hukuhara and generalized Hukuhara differences for fuzzy numbers [30, 31].

Let $f(x) : (a, b) \to T$ and $x_0 \in (a, b)$, then $f(x)$ is strongly differentiable at x_0 if there exists an element $f^{/}(x_0) \in T$ such that in the first form, if for $h > 0$ sufficiently near to zero there exist the H-differences $f(x_0 + h) \ominus f(x_0)$ and $f(x_0) \ominus f(x_0 - h)$.

$$f'(x) = \lim_{h \to 0^-} \frac{f(x_0 + h) \ominus f(x_0)}{h} = \lim_{h \to 0^-} \frac{f(x_0) \ominus f(x_0 - h)}{h}, \tag{1}$$

or also $f(x)$ is differentiable at x_0, in the second form, if for $h > 0$, sufficiently near 0, there exist the H-differences, $f(x_0) \ominus f(x_0 + h)$ and $f(x_0 - h) \ominus f(x_0)$.

$$f'(x) = \lim_{h \to 0^-} \frac{f(x_0) \ominus f(x_0 + h)}{-h} = \lim_{h \to 0^-} \frac{f(x_0 - h) \ominus f(x_0)}{-h}, \tag{2}$$

Currently, we can define an improper fuzzy function which is called Riemann integrable fuzzy function [32] for every $b \geq a$ *on* $[a, b]$. Let $f(x, r) = \left(\underline{f}(x, r), \overline{f}(x, r)\right)$ be a fuzzy-valued function on $[a, \infty]$, for every $b \geq a$ there exist two positive numbers; $\underline{T}(r)$ and $\overline{T}(r)$ such that $\int_a^b |\underline{f}(x, r)| dx \leq \underline{T}(r)$ and $\int_a^b |\overline{f}(x, r)| dx \leq \overline{T}(r)$, then $f(x)$ is improper fuzzy Riemann-integrable on $[a, \infty]$ and the improper integral for this is expressed as

$$I_u = \int_a^\infty u(x, r) dx = \int_a^\infty \underline{u}(x, r) dx, \int_a^\infty \overline{u}(x, r) dx. \tag{3}$$

Note that in this chapter, we signify the space of all continuous fuzzy functions on $[a, b] \subseteq \mathbb{R}$ and the space of all Lebesgue integrable fuzzy functions on the bounded $[a, b]$ by $\mathbb{C}^{\mathcal{F}}[a, b]$ and $\mathbb{L}^{\mathcal{F}}[a, b]$, respectively.

1.2 Fractional calculus on fuzzy numbers

A useful technique for solving problems in the real world that involve non-local operators is fractional calculus. Fuzzy fractional operators are a suitable alternative to represent the problems if the information and data are ambiguous. Fractional differentiability on fuzzy number-valued functions comes in various forms; a few are covered below.

Definition 1.1. [8, 33]

Let us consider $f(x, r)$ be a fuzzy number-valued function, for $(0 < r < 1)$ then Riemann-Liouville fractional derivative operator can be defined as

$$D_{a+}^{\alpha}\{f(x)\} = \frac{1}{\Gamma(m - \alpha)} \odot \left(\frac{d}{dx}\right)^m \int_a^x (x - u)^{m - \alpha - 1} \odot f(u) du, (x > a; \Re(\alpha) + 1 > 0 \tag{4}$$

$$D_{a-}^{\alpha}\{f(x)\} = \frac{1}{\Gamma(m - \alpha)} \odot \left(-\frac{d}{dx}\right)^m \int_x^b (u - x)^{m - \alpha - 1} \odot f(u) du, (x < b; \Re(\alpha) + 1 > 0) \tag{5}$$

These are left-sided and right-sided fractional derivatives of order α and $\Re(\alpha) > 0$.

We can say that $f(x)$ be a fuzzy-valued function and it is Riemann-Liouville H-differentiable about order $0 < \alpha < 1$ at any point $x_0 \in (a, b)$. According to the definition described by Laymaan [34], $f(x)$ be right and left differentiable fuzzy-valued function then:

$$D_{a+}^{\alpha}\{f(x_0)\} = \left[D_{a+}^{\alpha}\left(\underline{f}(x_0, r), D_{a+}^{\alpha}(\overline{f}(x_0, r)\right)\right],$$

and

$$D_{a+}^{\alpha}\{f(x_0)\} = \left[D_{a+}^{\alpha}(\overline{f}(x_0, r), D_{a+}^{\alpha}\left(\underline{f}(x_0, r)\right)\right], for\ 0 < r < 1$$

where,

$$\begin{cases} D^{\alpha}_{a+}\{\underline{f}\,(x_0,r)\} = \left[\dfrac{1}{\Gamma(1-\alpha)}\odot\dfrac{d}{dx}\displaystyle\int_a^x (x-u)^{-\alpha}\odot\underline{f}\,(u_0,r)du\right]_{x=x_0}, \\ D^{\alpha}_{a+}\{\bar{f}\,(x_0,r)\} = \left[\dfrac{1}{\Gamma(1-\alpha)}\odot\dfrac{d}{dx}\displaystyle\int_a^x (x-u)^{-\alpha}\odot\bar{f}(u_0,r)du\right]_{x=x_0}. \end{cases} \tag{6}$$

Definition 1.2. [31, 35].

The Riemann-Liouville fractional integral operator of order $\alpha > 0$ of a fuzzy-valued function $f(x,u)$ is defined as

$$I^{\alpha}_{a+}\{\mathrm{f}\,(\mathrm{x})\} = \frac{1}{\Gamma(\alpha)}\odot\int_a^x (x-u)^{\alpha-1}\odot f(u)du,\ (x>a;\Re(\alpha)>0) \tag{7}$$

and

$$I^{\alpha}_{b-}\{\mathrm{f}\,(\mathrm{x})\} = \frac{1}{\Gamma(\alpha)}\odot\int_x^b (u-x)^{\alpha-1}\odot f(u)du,\ (x<b;\Re(\alpha)>0) \tag{8}$$

I^{α}_{a+} and I^{α}_{b-} are the left-sided and right-sided fractional integrals of order α.

Particularly, $I^0\mathrm{f}\,(\mathrm{x}) = \mathrm{f}(\mathrm{x})$.

The fuzzy Riemann-Liouville integral of a fuzzy-valued function f can be expressed as follows:

$$I^{\alpha}_{a+}\{\mathrm{f}\,(\mathrm{x},\mathrm{r})\} = \left[I^{\alpha}_{a+}\left\{\underline{\mathrm{f}}\,(\mathrm{x},\mathrm{r})\right\}, I^{\alpha}_{a+}\{\bar{\mathrm{f}}\,(\mathrm{x}),\mathrm{r}\}\right], 0\le r\le 1$$

where

$$\begin{cases} I^{\alpha}_{a+}\left\{\underline{f}\,(x,r)\right\} = \dfrac{1}{\Gamma(\alpha)}\odot\displaystyle\int_a^x (x-u)^{\alpha-1}\odot\underline{f}(u)du, \\ I^{\alpha}_{a+}\{\bar{f}\,(x,r)\} = \dfrac{1}{\Gamma(\alpha)}\odot\displaystyle\int_a^x (x-u)^{\alpha-1}\odot\bar{f}(u)du. \end{cases} \tag{9}$$

2. Fuzzy fractional integral transform

We intend to define in this section, fuzzy integral transforms for functions with fuzzy values. Moreover, we will consider Laplace, Mellin and Sumudu fractional integral transforms with the properties of the fuzzy integral transforms.

2.1 Fuzzy fractional Laplace integral transform

Let $f\,(x)$ be a continuous fuzzy-value function. Suppose that $f\,(x)\odot e^{-tx}$ improper fuzzy Riemann integrable on $(0,\infty)$,

$$\mathcal{L}_f[f\,(x)] = \int_0^\infty f\,(x)\odot e^{-tx}dx = F(t),\ (t>0 \text{ and integer}) \tag{10}$$

In practice, the fuzzy Laplace transform approach is the most significant operational technique [22, 35]. Similarly, its significance is also defined and applicable in fractional calculus [18, 26, 36]. According to the procedure explained in Refs. [37–40], the solution of fuzzy fractional differential equations by using the fuzzy Laplace transform method as this approach, which is perhaps the most significant operative approach, also has the benefit of solving problems directly without requiring the development of nonhomogeneous differential equations in step two and the determination of a general solution in step one.

According to the derivative theorem to describe Laplace fractional derivatives scheme are:

Suppose that for a fuzzy-valued function, $f \in \mathcal{C}^{\mathcal{F}}[a,b] \cap \mathbb{L}^{\mathcal{F}}[a,b]$. Then:

$$\mathcal{L}_f\{D_{+0}{}^{\alpha} f(x,r)\} = t^{\alpha} F(t) \ominus D_{+0}{}^{\alpha-1} f(0),$$

where if $f(x)$ is $[(i)-\alpha]$ differentiable function and

$$\mathcal{L}_f\{D_{+0}{}^{\alpha} f(x,r)\} = -D_{+0}{}^{\alpha-1} f(0) \ominus - t^{\alpha} F(t),$$

where if $f(x)$ is $[(ii)-\alpha]$ differentiable function.

2.2 Fuzzy fractional Mellin integral transform

Let f be a continuous complex fuzzy-valued function [36]. Suppose that $x^{t-1} \odot f(x)$ is improper fuzzy Riemann-integrable on $(0,\infty)$ where $t \in \mathbb{C}$ then $\int_0^{\infty} x^{t-1} \odot f(x) dx$ is called fuzzy Mellin transforms and denoted by:

$$\mathcal{M}_f\{f(x),t\} = f(t) = \int_0^{\infty} x^{t-1} \odot f(x) dx, \tag{11}$$

Using Eq. (3), we have for $(0 \le r \le 1)$:

$$\int_0^{\infty} x^{t-1} \odot f(x,r) dx = \left[\int_0^{\infty} x^{t-1} \odot \underline{f}(x,r) dx. \int_0^{\infty} x^{t-1} \odot \bar{f}(x,r) dx \right], \tag{12}$$

Also, by using the definition of classical Mellin transform for fuzzy-valued function:

$$m\left\{\underline{f}(x,r)\right\} = \int_0^{\infty} x^{t-1} \odot \underline{f}(x,r) dx \text{ and} m\{\bar{f}(x,r)\} = \int_0^{\infty} x^{t-1} \odot \bar{f}(x,r) dx,$$

Then, we get

$$\mathcal{M}_f\{f(x,r),t\} = \left[m\left\{\underline{f}(x,r)\right\}, m\{\bar{f}(x,r)\} \right], \tag{13}$$

Since the fuzzy Mellin transform converges absolutely and uniformly in the vertical strip $(\alpha < \mathrm{Re}(s) < \beta)$ the inverse Mellin transform can be defined as

$$\mathcal{M}_f^{-1}\{f(t)\} = f(x) = \frac{1}{2\pi i} \int_{c-i\infty}^{c+i\infty} \mathcal{M}_f[f(x),t] x^{-t} dt, \tag{14}$$

where $\delta_- < c < \delta_+$,

Fractional fuzzy Mellin integral transform with all its operational properties is discussed by Azhar and Iqbal [41]; however, the same assumptions are applied for the solution of fuzzy fractional differential equations in the result section. Particularly, the derivative property of fuzzy Mellin transform used in the problem described by Hassan and Abbas [42] is as follows

$$\mathcal{M}_f\{D_x{}^{\alpha} f(x,r)\} = \frac{\Gamma(1-t+\alpha)}{\Gamma(1-t)} \odot f(t-\alpha), \tag{15}$$

where $f(t)$ is the fuzzy Mellin transform of the function $f(x,r)$.

2.3 Fuzzy fractional Sumudu integral transform

The fuzzy fractional Sumudu transform is an effective technique that enables the analysis and resolution of complex systems with uncertainty, and this transform offers a means of addressing the complexities of real-world situations where conventional approaches might not be sufficient by introducing the classical notions into the fuzzy domain.

Let for fuzzy-valued function which is continuous and let $e^{-x} \odot f(ux)$ is improper fuzzy Riemann-integrable on $(0,\infty)$where $s \in \mathbb{C}$ [42],

$$\mathcal{S}_f\{f(x),u\} = G(u) = \int_0^{\infty} e^{-x} \odot f(ux)dx, \text{ where } u \in [-\mu_1, \mu_2] \tag{16}$$

where the variable x in the fuzzy function's expression is factored using the variable u and $-\mu_1, \mu_2 > 0$. The parametric form of the fuzzy Sumudu transform is written as

$$\mathcal{S}_f\{f(x,r),u\} = \left[s\left\{\underline{f}(ux,r)\right\}, s\{\bar{f}(ux,r)\}\right], \tag{17}$$

where $s\left\{\underline{f}(x,r)\right\} = \int_0^{\infty} e^{-x} \odot \underline{f}(ux,r)dx$ and $s\{\bar{f}(x,r)\} = \int_0^{\infty} e^{-x} \odot \bar{f}(ux,r)dx$.

Fuzzy Sumudu and fuzzy Laplace transform are linked together using duality property for finding different useful results, expressed as

$$G\left(\frac{1}{x}\right) = xF(t), F\left(\frac{1}{u}\right) = uG(u). \tag{18}$$

Specifically, the fuzzy Sumudu transforms Riemann-Liouville H-differentiability property. Let f be the fuzzy-valued function with the relation

$$\mathcal{S}_f\{D_x{}^{\alpha} f(x,r)\}u = u^{-\alpha} G(u) \ominus \sum_{k=0}^{n-1} u^k D_{+0}{}^{\alpha-k-1} f(0). 0 \le r \le 1 \tag{19}$$

3. Fuzzy fractional differential equation

Fuzzy fractional order differential equations (FFDEs) are simply fuzzy differential equations (FDEs) that employ fractional operators. It is useful in providing an approximate solution for uncertain fractional models in a variety of physical or natural situations. Here, we consider fuzzy fractional differential equations with fuzzy initial-value problem,

$$D_x{}^{\alpha} f(x) = f(x, y(x)), y(x_0) = y_0 \tag{20}$$

where $f : [x_0, X] \times \mathbb{R} \rightarrow \mathbb{R}$ is a continuous mapping, $y_0, \in M$ is a fuzzy number with r-level intervals $[y_0]^r = \left[\underline{y_0}^r, \bar{y_0}^r\right]$, and $(0 \leq r \leq 1)$, and also we know that $f \in \mathbb{C}^{\mathcal{F}}[a, b] \cap \mathbb{L}^{\mathcal{F}}[a, b]$ be a fuzzy function. Here D_x^{α} is the Riemann-Liouville fractional derivative of non-integer α order. Now, mapping is defined for $y : [x_0, X] \rightarrow M$ be the solution of fuzzy differential equation if it is absolutely continuous and satisfies it everywhere on $[x_0, X]$. According to Zadeh's extension principle, when $y = y(x)$ is a fuzzy number.

$$f(x, y(x))(s) = \sup\{y(t) : s = f(x, t)\}, \quad t \in R$$

It follows that

$$[f(x, y(x))]^r = \left[\underline{f}^r(x, y(x)), \bar{f}^r(x, y(x))\right], \quad r \in [0, 1]$$

where,

$$\underline{f}^r(x, y(x)) = \min\left\{f(x, \alpha, u) : u \in \left[\underline{y}^r(x), \bar{y}^r(x)\right]\right\}, 0 < r < 1$$

and

$$\bar{f}^r(x, y(x)) = \max\left\{f(x, \alpha, u) : u \in \left[\underline{y}^r(x), \bar{y}^r(x)\right]\right\}, 0 < r < 1$$

Also, Ahmed et al. used Zadeh's extension principle to demonstrate the existence and uniqueness of analytical solutions to a fuzzy fractional initial-value problem [37].

3.1 Determination for solution of FFDEs by fuzzy Laplace transform

Consider Fuzzy fractional order differential equation with fuzzy initial value,

$$D_{+0}{}^{\alpha} f(x) = A(x) f(x, y(x)), D_{+0}{}^{\alpha-1}[y(x_0)] = y_0, x_0 \in (a, b) \tag{21}$$

where $A(x)$ is a fuzzy function and $B(x)$ is a given fuzzy input.

In this subsection, we provide the fuzzy Laplace transform and its inverse to derive solutions of FFDEs (Eq. (21)). Now applying Laplace transform on both sides of Eq. (21), we get

$$\mathcal{L}_f\{D_{+0}{}^{\alpha} f(x)\} = \mathcal{L}_f\{A(x) f(x, y(x))\}, \tag{22}$$

Afterward, we have the following cases according to the types of Riemann-Liouville H-differentiability:

Case I. If $y(x)$ is $[(i)-\alpha]$ differentiable function, then Eq. (22) is extended based on its lower and upper functions as follows:

$$\begin{cases} l\left\{D_{+0}{}^{\alpha}\underline{f}(x)\right\} = t^{\alpha} l\left\{\underline{y}(x,r)\right\} \ominus D_{+0}{}^{\alpha-1}\left(\underline{y}(0,r)\right), 0 \le r \le 1 \\ l\{D_{+0}{}^{\alpha}\bar{f}(x)\} = t^{\alpha} l\{\bar{y}(x,r)\} \ominus D_{+0}{}^{\alpha-1}(\bar{y}(0,r)), 0 \le r \le 1 \end{cases} \quad (23)$$

where $\underline{f}(x)$ is the minimum, while $\bar{f}(x)$ is the maximum function in the fuzzy domain.

$$\underline{f}(x,y(x);r) = \min\left\{f(x,t)|t \in \left(\underline{y}(x,r), \bar{y}(x,r)\right\}$$

and

$$\bar{f}(x,y(x);r) = \max\left\{f(x,t)|t \in \left(\underline{y}(x,r), \bar{y}(x,r)\right\}$$

For the linear system of equations, we have assumed that the solution of Eq. (23) as follows:

$$\begin{cases} l\left\{\underline{y}(x,r)\right\} = Q_1(t,r), \\ l\{\bar{y}(x,r)\} = R_1(t,r), \end{cases} \quad (24)$$

where $Q_1(t,r)$ and $R_1(t,r)$ are the solutions of the above system of equations, now by applying the inverse Laplace transform, we have solutions for $\underline{y}(x,r)$ and $\bar{y}(x,r)$:

$$\begin{cases} \underline{y}(x,r) = l^{-1}[Q_1(t,r)], \\ \bar{y}(x,r) = l^{-1}[R_1(t,r)]. \end{cases} \quad (25)$$

Case II. If $y(x)$ is $[(ii)-\alpha]$ differentiable function, then Eq. (22) we have

$$\begin{cases} l\left\{D_{+0}{}^{\alpha}\underline{f}(x)\right\} = t^{\alpha} l\left\{\underline{y}(x,r)\right\} \ominus D_{+0}{}^{\alpha-1}\left(\underline{y}(0,r)\right), 0 \le r \le 1 \\ l\{D_{+0}{}^{\alpha}\bar{f}(x)\} = t^{\alpha} l\{\bar{y}(x,r)\} \ominus D_{+0}{}^{\alpha-1}(\bar{y}(0,r)), 0 \le r \le 1 \end{cases} \quad (26)$$

For the linear system of equations, we have assumed that the solution of Eq. (26) is as follows:

$$\begin{cases} l\left\{\underline{y}(x,r)\right\} = Q_2(t,r) \\ l\{\bar{y}(x,r)\} = R_2(t,r), \end{cases} \quad (27)$$

where $Q_2(t,r)$ and $R_1(t,r)$ are the solutions of the above system of equations, now by applying the inverse Laplace transform, we have solutions for $\underline{y}(x,r)$ and $\bar{y}(x,r)$:

$$\begin{cases} \underline{y}(x,r) = l^{-1}[Q_1(t,r)], \\ \bar{y}(x,r) = l^{-1}[R_2(t,r)]. \end{cases} \quad (28)$$

3.2 General procedure for the solution of FFDEs by fuzzy Mellin transform

Consider a fuzzy fractional order differential equation with fuzzy initial value,

$$D_{+0}{}^{\alpha}f(x) = f(x, y(x)), D^{-\alpha}\{y(x_0)\} = y_0 \tag{29}$$

where $D_{+0}{}^{\alpha}$ be the left-sided Riemann-Liouville fractional derivative. Now applying the Mellin transform on above, we have:

$$\mathcal{M}_f\{D_{+0}{}^{\alpha}f(x)\} = \mathcal{M}_f\{f(x, y(x))\}, \tag{30}$$

We then obtain by recalling [8] for $\Re(t) < \Re(\alpha) + 1$

$$\sum_{k=0}^{m} A^k \frac{\Gamma(1-t)}{\Gamma(1-t-\alpha-k)} \odot \mathcal{M}_f y(t) = \mathcal{M}_f\{f(x)\}, \tag{31}$$

Since f be a fuzzy-valued function, then above it contains lower and upper function as $f(x, r) = \left\{\underline{f}(x, r), \bar{f}(x, r)\right\}$, $(0 < r < 1)$

Now by applying the inverse Mellin transform the following solution can be derived,

$$y(x) = \mathcal{M}_f{}^{-1}\left\{\frac{1}{Q_\alpha(1-t)} \odot \mathcal{M}_f f(t)\right\}, \tag{32}$$

where $Q_\alpha(1-t) = \sum_{k=0}^{m} A^k \frac{\Gamma(t)}{\Gamma(t-\alpha-k)}$, or

$$\begin{cases} \underline{y}(x) = \mathcal{M}_f{}^{-1}\left\{\frac{1}{Q_\alpha(1-t)} \odot \mathcal{M}_f\left\{\underline{f}(t)\right\}\right\}, \\ \bar{y}(x) = \mathcal{M}_f{}^{-1}\left\{\frac{1}{Q_\alpha(1-t)} \odot \mathcal{M}_f\left\{\bar{f}(t)\right\}\right\}. \end{cases} \tag{33}$$

Also, we can write it in the form of Mellin fractional analog of Green's function as

$$G_\alpha(t) = \mathcal{M}_f{}^{-1}\left\{\frac{1}{Q_\alpha(1-t)}(t)\right\}, \tag{34}$$

Solutions can be obtained by applying the property of Mellin convolution.

$$\mathcal{M}_f\left\{\int_0^\infty f(\xi) g\left(\frac{x}{\xi}\right)\frac{d\xi}{\xi}\right\} = \mathcal{M}_f\{(f(x)\} \odot \mathcal{M}_f\{g(x)\}, \tag{35}$$

We get

$$y(x) = \int_0^\infty G_\alpha\left(\frac{x}{\xi}\right) \odot f(\xi) \odot \frac{d\xi}{\xi}, \tag{36}$$

Note: The only difference between left- and right-sided Riemann-Liouville fractional operators is in the initial points, hence in our explanation, we are utilizing the right-sided Riemann-Liouville fractional derivative operator as a practical operator.

3.3 General procedure for the solution of FFDEs by fuzzy Sumudu transform

Consider the following FFDEs

$$D_{+0}{}^{\alpha}f(x) = f(x, y(x)), D_{+0}{}^{\alpha-1}[y(x_0)] = y_0, \tag{37}$$

Let $f \in C^{\mathcal{F}}[a,b] \cap \mathbb{L}^{\mathcal{F}}[a,b]$ be a fuzzy function,
Now applying the Sumudu transform on both sides of Eq. (37), we get

$$\mathcal{S}_f\{D_{+0}{}^{\alpha}f(x)\} = \mathcal{S}_f\{f(x, y(x))\}, \tag{38}$$

Then, we have the following cases according to the types of Riemann-Liouville H-differentiability:

Case I. If $y(x)$ is $[(i) - \alpha]$ differentiable function, then Eq. (38) is extended based on its lower and upper functions as follows:

$$\begin{cases} s\left\{D_{+0}{}^{\alpha}\underline{f}(x)\right\} = u^{-\alpha}s\left\{\underline{y}(x,r)\right\} \ominus D_{+0}{}^{\alpha-1}\left(\underline{y}(0,r)\right), 0 \le r \le 1 \\ s\{D_{+0}{}^{\alpha}\bar{f}(x)\} = u^{-\alpha}s\{\bar{y}(x,r)\} \ominus D_{+0}{}^{\alpha-1}(\bar{y}(0,r)), 0 \le r \le 1 \end{cases} \tag{39}$$

where $\underline{f}(x)$ be the minimum while $\bar{f}(x)$ be the maximum function in the fuzzy domain.

$$\underline{f}(x, y(x); r) = \min\left\{f(x,u) | u \in \left(\underline{y}(x,r), \bar{y}(x,r)\right)\right\}$$

and

$$\bar{f}(x, y(x); r) = \max\left\{f(x,u) | u \in \left(\underline{y}(x,r), \bar{y}(x,r)\right)\right\}$$

For the linear system of equations, we have the solution of Eq. (39) as follows:

$$\begin{cases} (1-u^{\alpha})s\left\{\underline{y}(x,r)\right\} = \underline{y_0}(0,r)u^{\alpha}, \\ (1-u^{\alpha})s\{\bar{y}(x,r)\} = \bar{y}(0,r)u^{\alpha}. \end{cases} \tag{40}$$

Now by applying the inverse Sumudu transform, we have solutions for $\underline{y}(x,r)$ and $\bar{y}(x,r)$:

$$\begin{cases} \underline{y}(x,r) = \underline{y_0}(0,r)\, s^{-1}\left[\dfrac{u^{\alpha}}{1-u^{\alpha}}\right], \\ \bar{y}(x,r) = \bar{y}(0,r)\, s^{-1}\left[\dfrac{u^{\alpha}}{1-u^{\alpha}}\right]. \end{cases} \tag{41}$$

Case II. If $y(x)$ is $[(ii) - \alpha]$ differentiable function, then from Eq. (38), we have

$$\begin{cases} s\left\{D_{+0}{}^{\alpha}\underline{f}(x)\right\} = -D_{+0}{}^{\alpha-1}\left(\underline{y}(0,r)\right) \oplus u^{-\alpha}s\left\{\underline{y}(x,r)\right\}, 0 \le r \le 1 \\ s\{D_{+0}{}^{\alpha}\bar{f}(x)\} = -D_{+0}{}^{\alpha-1}(\bar{y}(0,r)) \oplus u^{-\alpha}s\{\bar{y}(x,r)\}, 0 \le r \le 1 \end{cases} \tag{42}$$

For the linear system of equations, we have the solution of Eq. (42) as follows:

$$\begin{cases} s\left\{D_{+0}{}^{\alpha}\underline{f}(x)\right\} = -D_{+0}{}^{\alpha-1}\left(\underline{y}(0,r)\right)\oplus u^{-\alpha}s\left\{\underline{y}(x,r)\right\}, 0\leq r\leq 1 \\ s\{D_{+0}{}^{\alpha}\bar{f}(x)\} = -D_{+0}{}^{\alpha-1}(\bar{y}(0,r))\oplus u^{-\alpha}s\{\bar{y}(x,r)\}, 0\leq r\leq 1 \end{cases} \tag{43}$$

Then by applying the inverse Laplace transform, we have solutions for $\underline{y}(x,r)$ and $\bar{y}(x,r)$

$$\begin{cases} (1-u^{\alpha})s\left\{\underline{y}(x,r)\right\} = \underline{y_0}(0,r)u^{\alpha}, \\ (1-u^{\alpha})s\{\bar{y}(x,r)\} = \bar{y}(0,r)u^{\alpha}. \end{cases} \tag{44}$$

4. Applications with examples

Example 1.

Now, we consider the fuzzy differential equation.

$$t^{\alpha+1}\odot D^{\alpha+1}y(t) + \tau\odot t^{\alpha}\odot D^{\alpha}y(t) = f(t), \text{for } t>0; \alpha>0; \tau\in\mathbb{R} \tag{45}$$

Then particular solutions of the above fuzzy fractional order differential equation are different for values of τ.

Case I: For $\alpha+\tau\neq n$ with $n\in\mathbb{N}$

If $\alpha>0$, $y=y(t,r)$ be a fuzzy-valued function. Then Eq. (45) is solvable, and its particular solution is given by

$$y(t,r) = \int_t^{\infty} G_{\alpha,\tau}\left(\frac{t}{\xi}\right)\odot f(\xi)\odot\frac{d\xi}{\xi}, 0<r<1 \tag{46}$$

where $G_{\alpha}(t)$ is the Mellin fractional analog of Green's function.

$$G_{\alpha,\tau}(t) = G_{\alpha}(t) = \mathcal{M}_f{}^{-1}\left\{\frac{\Gamma(s)}{(s+\alpha+\tau)\Gamma(s+\alpha)}\right\}(t), \tag{47}$$

Now in accordance with the convolution property of the Mellin transform, we can consider.

$$\mathcal{M}_f\{G(s)\} = \mathcal{M}_f\{G_1(s)\}\odot\mathcal{M}_f\{G_2(s)\},$$

where

$$\mathcal{M}_f\{G_1(s)\} = \frac{\Gamma(s)}{\Gamma(s+\alpha)}, \text{and } \mathcal{M}_f\{G_2(s)\} = \frac{1}{(s+\alpha+\tau)} \tag{48}$$

The inverse of the Mellin transform of Eq. (36) gives the results.

$$G_{1,\alpha}(t) = G_{2,\alpha}(t) = 0, \text{for } t>1$$

and

$$G_{1,\alpha}(t) = \frac{(t)^{\alpha-1}}{\Gamma(\alpha)} \text{ and } G_{2,\alpha}(t) = t^{\alpha+\tau}, for\ 0<t<1 \tag{49}$$

Thus, we can express it in the form of Mellin convolution in Eq. (46)

$$G_{\alpha,\tau}(t) = \left\{ \int_0^\infty G_{1,\alpha}\left(\frac{t}{\xi}\right) \odot G_{2,\alpha}(\xi) \frac{d\xi}{\xi} \right\}, \tag{50}$$

The residue theorem can be used to find the solution to Eq. (36) as the definition of fuzzy inverse Mellin transforms into an explicit representation also demonstrates.

$$G_\alpha(t) = \frac{1}{2\pi i} \int_{\gamma - i\infty}^{\gamma + i\infty} \frac{\Gamma(s)}{(s + \alpha + \tau)\Gamma(s + \alpha)} \odot t^{-s} \odot ds, \tag{51}$$

Since, $\alpha + \tau \neq n(n \in \mathbb{N})$, it has a pole at $s = -\alpha - \tau$ of the integrand of Eq. (51), which does not coincide with any pole at $s = -k\ (k \in \mathbb{N})$ of $\Gamma(s)$, which provide the solution for $\gamma > max\,[-\alpha - \Re(\tau), 0]$ and the evaluation of residues yields:

$$G_\alpha(t) = \frac{\Gamma(-\alpha - \tau) \odot t^{\alpha+\tau}}{\Gamma(-\tau)} \ominus \sum_{k=0}^{\infty} \frac{(-1)^k \odot \Gamma(k - \tau - \alpha) \odot t^k}{k! \odot \Gamma(k + 1 - \tau - \alpha) \odot \Gamma(\alpha - k)}, \tag{52}$$

Corollary: Let $\alpha = \frac{1}{2}$ in Eq. (34) *for* $\tau \neq n - 1/2$ then

$$t^{\frac{3}{2}} \odot D^{\frac{3}{2}} y(t) + \tau \odot t^{\frac{1}{2}} \odot D^{\frac{1}{2}} y(t) = f(t), \tag{53}$$

has the solution of the form,

$$y(t, r) = \int_t^\infty G_{1/2,\tau}\left(\frac{t}{\xi}\right) \odot f(\xi) \odot \frac{d\xi}{\xi}, 0 < r < 1 \tag{54}$$

where Eq. (41) becomes for $\alpha = \frac{1}{2}$,

$$G_{1_{\frac{1}{2}}}(t) = \Gamma(-½ - \tau) \odot \frac{t^{½+\tau}}{\Gamma(-\tau)} \ominus \sum_{k=0}^{\infty} (-1)^k \odot \Gamma\left(\frac{k - \tau - ½) \odot t^k}{k! \odot \Gamma(k + 1 - \tau - ½) \odot \Gamma(½ - k)}\right., \tag{55}$$

Case II: According to Kilbas et al. [8], the fractional order differential equation Eq. (45) is solvable if $\alpha > 0$ and $n \in N$ are such that $\alpha \neq 1, 2, \ldots, n$. Its particular solution can be found in Eq. (46). The reason behind this is that the function $G_{\alpha,n-\alpha}(x)$ is defined for $\tau = n - \alpha$ such that $\alpha = k\ (k = 1, 2, \ldots, n)$.
where

$$G_{\alpha,n-\alpha}(x) = \frac{(-1)^{n-1} \odot t^n}{n! \Gamma(\alpha - n)} \left[logt + \sum_{i=0}^{n-1} \frac{1}{i - n} + \psi(\alpha - n) + \gamma \right] + \sum_{\substack{k=0 \\ k \neq n}}^{\infty} \frac{(-1)^k \odot t^k}{k! \odot \Gamma(n - k) \odot \Gamma(\alpha - k)}, \tag{56}$$

Eq. (56) is the particular solution of the FFDE by Fuzzy Mellin transform method, is given by

$$t^{\alpha+1} \odot D^{\alpha+1} y(t) + (n - \alpha) \odot t^\alpha \odot D^\alpha y(t) = f(t). \tag{57}$$

Example 2.

Let us find the solution of fractional fuzzy differential equation with fuzzy initial value,

$$D_{+0}{}^{\alpha}f(x) = \mu \odot y(x), D_{+0}{}^{\alpha-1}[y(x_0)] = y_0, x_0 \in (a,b) \tag{58}$$

where $y(x)$ is the number of radionuclides present in a given radioactive, μ is a decay constant. This example can be solved according to two following cases for $\mu \in \mathbb{R}$.

Now applying Laplace transform on both sides of Eq. (13), we get

$$\mathcal{L}_f\{D_{+0}{}^{\alpha}f(x)\} = \mathcal{L}_f\{\mu \odot y(x)\} \tag{59}$$

Now, we have the following cases according to the types of Riemann-Liouville H-differentiability:

Case I. If $y(x)$ is $[(i)-\alpha]$ differentiable function, let $\mu \in \mathbb{R}^+ = (0, +\infty)$then Eq. (59) is extended based on its lower and upper functions as follows:

$$\begin{cases} \mu l\left\{\underline{y}(x,r)\right\} = t^{\alpha} l\left\{\underline{y}(x,r)\right\} \ominus D_{+0}{}^{\alpha-1}\left(\underline{y}(0,r)\right), 0 \le r \le 1 \\ \mu l\{\bar{y}(x,r)\} = t^{\alpha} l\{\bar{y}(x,r)\} \ominus D_{+0}{}^{\alpha-1}(\bar{y}(0,r)), 0 \le r \le 1 \end{cases} \tag{60}$$

For the solution of a linear system of equations Eq. (60), with some simplifications and operations are as follows:

$$\begin{cases} (t^{\alpha} - \mu) l\left\{\underline{y}(x,r)\right\} = D_{+0}{}^{\alpha-1}\left(\underline{y}(0,r)\right), \\ (t^{\alpha} - \mu) l\{\bar{y}(x,r)\} = D_{+0}{}^{\alpha-1}(\bar{y}(0,r)), \end{cases} \tag{61}$$

Subsequently, now by applying the inverse Laplace transform, we have solutions for $\underline{y}(x,r)$ and $\bar{y}(x,r)$ are

$$\begin{cases} \underline{y}(x,r) = D_{+0}{}^{\alpha-1}\left(\underline{y}(0,r)\right) l^{-1}\left[\dfrac{1}{(t^{\alpha} - \mu)}\right], \\ \bar{y}(x,r) = D_{+0}{}^{\alpha-1}(\bar{y}(0,r)) l^{-1}\left[\dfrac{1}{(t^{\alpha} - \mu)}\right]. \end{cases} \tag{62}$$

Case II. Now for $\mu \in \mathbb{R} = (-\infty, 0)$ and $y(x)$ is $[(ii)-\alpha]$ differentiable function, then we have

$$\begin{cases} \mu l\left\{\underline{y}(x,r)\right\} = -D_{+0}{}^{\alpha-1}\left(\underline{y}(0,r)\right) \ominus - t^{\alpha} l\left\{\underline{y}(x,r)\right\}, 0 \le r \le 1 \\ \mu l\{\bar{y}(x,r)\} = -D_{+0}{}^{\alpha-1}(\bar{y}(0,r)) t^{\alpha} \ominus - t^{\alpha} l\{\bar{y}(x,r)\}, 0 \le r \le 1 \end{cases} \tag{63}$$

For the linear system of equations, we have assumed that the solution of Eq. (63) is as follows:

$$\begin{cases} (t^{\alpha} - \mu) l\left\{\underline{y}(x,r)\right\} = D_{+0}{}^{\alpha-1}\left(\underline{y}(0,r)\right), \\ (t^{\alpha} - \mu) l\{\bar{y}(x,r)\} = D_{+0}{}^{\alpha-1}(\bar{y}(0,r)), \end{cases} \tag{64}$$

Now by applying the inverse Laplace transform, we have solutions for $\underline{y}(x,r)$ and $\bar{y}(x,r)$

$$\begin{cases} \underline{y}(x,r) = D_{+0}{}^{\alpha-1}\left(\underline{y}(0,r)\right)l^{-1}\left[\dfrac{1}{(t^{\alpha}-\mu)}\right], \\ \bar{y}(x,r) = D_{+0}{}^{\alpha-1}(\bar{y}(0,r))l^{-1}\left[\dfrac{1}{(t^{\alpha}-\mu)}\right]. \end{cases} \tag{65}$$

Example 3.

Consider a fuzzy fractional order differential equation with a fuzzy initial value,

$$D_{+0}{}^{\alpha}f(x) = \lambda\odot y(x), D_{+0}{}^{\alpha-1}\left[y(x_0)\right] = y_0, x_0 \in (a,b) \tag{66}$$

Now applying the Sumudu transform on both sides of Eq. (66), we get

$$S_f\{D_{+0}{}^{\alpha}f(x)\} = S_f\{\lambda\odot y(x)\}, \tag{67}$$

Then, we have the following cases according to the types of Riemann-Liouville H-differentiability:

Using the derivative property of fuzzy Sumudu transform

$$S_f\{D_{+0}{}^{\alpha}f(x)\} = u^{\alpha}S_f\{f(x)\} - \sum_{k=0}^{n-1} u^{\alpha-1-k}f^{k}(0) \tag{68}$$

Case I. If $y(x)$ is $[(i)-\alpha]$ differentiable function, then Eq. (68) can be expressed particularly for k = 0 as,

$$\begin{cases} \lambda s\left\{\underline{y}(x,r)\right\} = u^{-\alpha}s\left\{\underline{y}(x,r)\right\}\ominus D_{+0}{}^{\alpha-1}\left(\underline{y}(0,r)\right), 0\le r\le 1 \\ \lambda s\{\bar{y}(x,r)\} = u^{-\alpha}s\{\bar{y}(x,r)\}\ominus D_{+0}{}^{\alpha-1}(\bar{y}(0,r)), 0\le r\le 1 \end{cases} \tag{69}$$

For the linear system of equations, we have the Eq. (69) as follows

$$\begin{cases} (1-\lambda u^{\alpha})s\left\{\underline{y}(x,r)\right\} = \underline{y_0}(0,r)u^{\alpha}, \\ (1-\lambda u^{\alpha}))s\{\bar{y}(x,r)\} = \bar{y}(0,r)u^{\alpha}. \end{cases} \tag{70}$$

Now by applying the inverse Sumudu transform, we have solutions for $\underline{y}(x,r)$ and $\bar{y}(x,r)$

$$\begin{cases} \underline{y}(x,r) = \underline{y_0}(0,r)\, s^{-1}\left[\dfrac{u^{\alpha}}{1-\lambda u^{\alpha})}\right], \\ \bar{y}(x,r) = \bar{y}(0,r)\, s^{-1}\left[\dfrac{u^{\alpha}}{1-\lambda u^{\alpha})}\right]. \end{cases} \tag{71}$$

Case II. If $y(x)$ is $[(ii)-\alpha]$ differentiable function, then Eq. (66), then we have

$$\begin{cases} \lambda s\left\{\underline{y}(x,r)\right\} = -D_{+0}{}^{\alpha-1}\left(\underline{y}(0,r)\right)\oplus u^{-\alpha}s\left\{\underline{y}(x,r)\right\}, 0\le r\le 1 \\ \lambda s\{\bar{y}(x,r)\} = -D_{+0}{}^{\alpha-1}(\bar{y}(0,r))\oplus u^{-\alpha}s\{\bar{y}(x,r)\}, 0\le r\le 1 \end{cases} \tag{72}$$

For the linear system of equations, we have the solution of Eq. (72) as follows:

$$\begin{aligned}(1-\lambda u^{\alpha})s\left\{\underline{y}(x,r)\right\} &= \underline{y_0}(0,r)u^{\alpha},\\ (1-\lambda u^{\alpha}))s\{\bar{y}(x,r)\} &= \bar{y}(0,r)u^{\alpha}.\end{aligned} \tag{73}$$

Then by applying the inverse Sumudu transform, we have solutions for $\underline{y}(x,r)$ and $\bar{y}(x,r)$

$$\begin{cases}\underline{y}(x,r) = \underline{y_0}(0,r)\, s^{-1}\left[\dfrac{u^{\alpha}}{1-\lambda u^{\alpha})}\right],\\ \bar{y}(x,r) = \bar{y}(0,r)\, s^{-1}\left[\dfrac{u^{\alpha}}{1-\lambda u^{\alpha})}\right].\end{cases} \tag{74}$$

5. Conclusion

All the descriptions in this chapter are concluded with the following facts:

- Through incorporating fuzzy logic and fractional calculus, the fuzzy fractional integral transform is an expansion of conventional integral transforms. It is helpful in resolving issues involving uncertainty and dynamics of non-integer order.

- The key concepts used in this chapter for describing the fuzzy fractional integral transformations are fuzzy logic, fractional calculus, and fuzzy integral transformations.

- We have found that it is an effective method for examining and resolving issues involving uncertainty and fractional dynamics. It unites the ideas of fractional calculus with fuzzy logic concepts to offer a solid foundation for handling complex real-world problems.

- In order to solve fuzzy fractional order differential equations, fuzzy integral transformations have been investigated.

- The solution of FFDEs is examined using fractional fuzzy Laplace, Mellin, and Sumudu integral transforms. In this chapter, under Riemann-Liouville fractional H-differentiability, we have examined the analytical solution of fuzzy fractional order differential equations, or FFDEs.

- Consequently, in order to study the solutions of FFDEs using the right-sided Riemann-Liouville fractional derivative, we have solved a few well-known cases in addition to the general technique in a fractional way.

- The lower and upper functions of the fuzzy solution are ascertained as follows from the general solution of the fuzzy fractional order differential equation.

- In the case of fractional fuzzy Mellin transform, we have

$$\begin{cases} \underline{y}(x,r) = \int_x^{\infty} g\left(\frac{x}{\xi}\right) \odot \underline{f}(\xi) \odot \frac{d\xi}{\xi}, & 0<r<1 \\ \bar{y}(x,r) = \int_x^{\infty} g\left(\frac{x}{\xi}\right) \odot \bar{f}(\xi) \odot \frac{d\xi}{\xi}, & 0<r<1 \end{cases} \tag{75}$$

- In the case of fuzzy fractional Laplace transform and Sumudu transform, the solution can be obtained by convolution property:

$$\begin{cases} \underline{y}(x,r) = \int_0^{x} g(x-\xi) \odot \underline{f}(\xi) \odot d\xi, & 0<r<1 \\ \bar{y}(x,r) = \int_0^{x} g(x-\xi) \odot \bar{f}(\xi) \odot d\xi, & 0<r<1 \end{cases} \tag{76}$$

- In order to discover an approximate solution with stability and error analysis, we can thus validate our approach to solving FFDEs using numerical approaches in future work.

Acknowledgements

The authors are grateful to the learned referee for a careful reading of the manuscript with many useful suggestions for improvement.

Author's contributions

The authors have contributed equally to the writing of this paper. They read and approved the manuscript.

Conflict of interests

The authors declare that they have no competing interests.

Availability of data and materials

The sharing of data does not apply to this article since no data sets were produced or analyzed during the present study.

Author details

Noreen Azhar[1]*, Saleem Iqbal[2], Raheela Manzoor[1] and Shazia Kalsoom[1]

1 Mathematics Department, Sardar Bahadur Khan Women's University, Quetta, Pakistan

2 Mathematics Department, University of Balochistan, Quetta, Pakistan

*Address all correspondence to: noreen_nadeem@yahoo.com

IntechOpen

© 2025 The Author(s). Licensee IntechOpen. This chapter is distributed under the terms of the Creative Commons Attribution License (http://creativecommons.org/licenses/by/4.0), which permits unrestricted use, distribution, and reproduction in any medium, provided the original work is properly cited.

References

[1] Agarwal RP, Lakkshmikantham V, Nieto JJ. On the concept of solution for fractional differential equations with uncertainty. Nonlinear Analysis. 2010; **72**:2859-2862

[2] Baglay RL. On the fractional order initial value problem and its engineering applications. In: Nishimoto K, editor. Fractional Calculus and Its Applications. Collage of Engineering. Tokoyo: Nihon University; 1990. pp. 12-200. [Proc. Int. Conf. held at Nihon Univ., Tokyo. 1989

[3] Beyer H, Kempfle S. Definition of physically consistent damping laws with fractional de-rivatives. Zamm-zeitchrift Fur Angewandte Mathematik Und Mechanik. 1995;**75**:623-635

[4] West BJ, Bologna M, Griolini P. Physics of Fractional Operators. Berlin: Springer-Verlag; 2003

[5] Magin RL. Fractional Calculus in Bioengineering. Redding: Begell House Publisher; 2006

[6] Oldham KB, Spanier J. Applications to diffusion problems. In: Fractional Calculus: Theory and Applications, Differentiation, and Integration into Arbitrary Order. New York: Academic Press; 1974

[7] Podlubny I. Fractional-order systems and fractional-order controllers. In: Fractional Differential Equations: An Introduction to Fractional Derivatives, Fractional Differential Equations, Some Methods of Their Solution and Some of Their Applications. San Diego: Academic Press; 1999

[8] Kilbas AA, Srivastava HM, Trujillo JJ. Theory and Applications of Fractional Differential Equations. 1st ed. Amsterdam: Elsevier Science B.V; 2006

[9] Mainardi F. Fractional Calculus and Waves in Linear Viscoelasticity: An Introduction to Mathematical Models. London: Imperial College Press; 2010

[10] Lakshmikantham V, Vatsala AS. Basic theory of fractional differential equations. Nonlinear Anal. 2008;**69**: 2677-2682

[11] Gorenflo R, Mainardi F. Random walk models for space-fractional diffusion processes. Fractional Calculus and Applied Analysis. 1998;**1**(2):167-190

[12] Mainardi F. Fractional relaxation-oscillation and fractional diffusion-wave phenomena. Chaos, Solitons and Fractals. 1996;**7**:1461-1477

[13] Podlubny I, editor. Fractional derivatives and integrals. In: Fractional Differential Equations. San Diego: Academic Press; 1999. pp. 41-119

[14] Belgacem FBM, Karballi AA. Sumudu transforms fundamental properties. Journal of Applied Mathematics and Stochastic Analysis. 2005;**2006**:1-23

[15] Kılıçman A, Altun O. Some remarks on the fractional Sumudu transform and applications. Applied Mathematics and Information Sciences. 2014;**8**(6): 2881-2888

[16] Asiru MA. Further properties of the Sumudu transform and its applications. International Journal of Mathematical Education in Science and Technology. 2002;**33**:441-449

[17] Kaleva O. Fuzzy differential equations. Fuzzy Sets and Systems. 1987; **24**:301-317

[18] Lupulescu V, Arshad S. Fractional differential equations with fuzzy initial

condition. Electronic Journal of Differential Equations. 2011; **2011**(34):1-8

[19] Buckley JJ, Feuring T. Fuzzy differential equations. Fuzzy Sets and Systems. 2000;**110**:43-54

[20] Hammouch Z, Dutta H, Melliani S, Ruzhansky M, editors. Existence and uniqueness results of fractional differential equations with fuzzy data. In: Nonlinear Analysis: Problems, Applications and Computational Methods. Cham: Springer; 2021

[21] Noeiaghdam S, Sidorov D. Caputo-Fabrizio fractional derivative to solve the fractional model of energy supply-demand system. Mathematical Modelling of Engineering Problems. 2020;7(3):359-367

[22] Ahmadi MB, Kiani NA, Mikaeilvand N. Laplace transform formula on fuzzy nth-order derivative and its application in fuzzy ordinary differential equations. Soft Computing. 2014;**18**: 2461-2469

[23] Gouyandeh Z, Allahviranloo T, Abbasbandy S, Armand A. A fuzzy solution of heat equation under generalized Hukuhara differentiability by fuzzy Fourier transform. Fuzzy Sets and Systems. 2017;**309**:81-97

[24] Zadeh LA. Fuzzy sets. Information and Control. 1965;**8**(3):338-353

[25] Harir A, Melliani S, Chadli LS. Fuzzy conformable fractional differential equations. International Journal of Differential Equations. 2021; **2021**:6

[26] Harir A, Melliani S, Chadli LS. Existence, Uniqueness and Approximate Solutions of Fuzzy Fractional Differential Equations. 2020

[27] Atimad H, Melliani S, Chadli LS. Fuzzy fractional evolution equations and fuzzy solution operators. Advances in Fuzzy Systems. 2019;**2019**:10

[28] Tofigh A, Noeiaghdam S, Noeiaghdam S, Nieto JJ. A fuzzy method for solving fuzzy fractional differential equations based on the generalized fuzzy Taylor expansion. Mathematics. 2020;**8** (12):1-24

[29] Bede B. Mathematics of Fuzzy Sets and Fuzzy Logic. Heidelberg: Springer; 2013

[30] Bede B, Stefanini L. Generalized differentiability of fuzzy-valued functions. Fuzzy Sets and Systems. 2013; **230**:119-141

[31] Podlubny I. Fractional Differential Equations. San Diego: Academic Press; 1999

[32] Stefanini L, Bede B. Generalized Hukuhara differentiability of interval-valued functions and interval differential equations. Nonlinear Analysis: Theory Methods & Applications. 2009;**71**:1311-1328

[33] Allahviranloo T. Fuzzy Fractional Differential Operators and Equations. Istanbul: Springer International Publishing; 2021. pp. 88-103

[34] Laymaan JW. The Hankel transform and some of its properties. J Integer Seq. 2001;**4**(1):1-11

[35] Jafarian A, Golmankhaneh AK, Baleanu D. On fuzzy fractional Laplace transformation. Advances in Mathematical Physics. 2014;**2014**:1-9

[36] Allahviranloo T, Salahhshour S, Abbasbandy S. Explicit solutions of fractional differential equations with

uncertainty. Soft Computing. 2012;**16**: 297-302

[37] Ahmad MZ, Hassan MK, Abbasbanday S. Solving fuzzy fractional differential equations using Zadeh's extension principle. Scientific World Journal. 2013;**2013**:1-11

[38] Arshad S, Lupulescu V. On the fractional differential equations with uncertainty. Nonlinear Analysis: Theory Methods & Applications. 2011;**74**(11): 3685-3693

[39] Mazandarani M, Kamyad AV. Modified fractional Euler method for solving fuzzy fractional initial value problem. Communications in Nonlinear Science and Numerical Simulation. 2013; **18**(1):12-21

[40] Kexue L, Jigen P. Laplace transform and fractional differential equations. Applied Mathematics Letters. 2011;**24** (12):2019-2023

[41] Azhar N, Iqbal S. Solution of fuzzy fractional differential equations by fractional Mellin transform method. Journal of Computational and Applied Mathematics. 2022;**200**:1-11

[42] Hassan RH, Abbas YM. Fuzzy Sumudu transforms of the fuzzy Riemann-Liouville fractional derivatives about order. International Journal of Engineering and Technology. 2018;**7**(4): 363-367

Chapter 6

A Review of the State-of-the-Art in Fractional-Order PID Controllers

Yunus Emre Yokuş, Umut Engin Ayten and Aslıhan Kartcı

Abstract

Fractional-order (FO) PID controllers are becoming increasingly popular for various industrial applications due to the advantages they can offer. The increasing tendency toward the utilization of FOPID controllers is mainly owing to the fact that these controllers have extra tuning knobs that allow coherent adjustment of the dynamics of control systems. Nonetheless, there are several critical points that remain underexplored in the existing literature, namely: (i) What are the significant advantages of FOPID controllers in comparison with integer-order (IO) PID controllers? (ii) For the real-time implementation of FOPID controllers, approximations equivalent to high-order linear controllers are used. In this case, what are the benefits of using FOPID controllers? Lastly, (iii) What are the advantages of having a near-ideal FO behavior in the control practice? In this chapter, we attempt to address these issues through conducting a comprehensive review of relevant publications in the rapidly growing literature on FO control, outlines the milestones and disadvantages, and presents future perspectives for the industrialization of FO control. Moreover, we draw conclusions and formulate suggestions for future searches. In conclusion, this chapter helps the beginners to get started quickly and to learn how to select, tune, approximate, and implement FOPID controllers in the frequency domain.

Keywords: fractional calculus, fractional-order control, fractional-order PID (FOPID) controller, industrial applications of FOPID, fractional control implementation

1. Introduction

An integer-order proportional-integral-derivative (IOPID) controller was first introduced in 1911 for use in automatic ship steering by Sperry [1]. IOPID has since maintained its relevance owing to its ease in design and implementation. In most process control industries, 95% of the control loops are of IOPID type because of several advantages [2]. But most of the failures encountered in industrial controllers are because of underperforming IOPID loops, which is due to poor tuning rules. Twenty-five percent of process control loops use factory default settings. Also, these control loops produce less variance in manual mode than with automatic, which is a serious situation in industries [3]. In a study, Wang has reported that only 20% of industrial control loops work properly [4].

In the IOPID controller, the derivative controller increases the relative stability, and the integral controller eliminates the steady-state error. However, the derivative action not only introduces $\pi/2$ phase lead to maintain the relative stability, but also increases the system gain and thereby sensitivity [5, 6]. In order to avoid the sensitivity issue of IOPID controllers, the majority of the real-time industries, including process control, refineries, chemical, pulp, and paper industries, have started deploying IOPI controllers for the regulatory actions, with derivative gain being forced to zero. In addition to the sensitivity problem due to the differentiator, the integrator in IOPID also introduces reduction in relative stability [5, 6]. This raises the question of whether or not to have an integrator/differentiator in the IOPID controller design. A satisfactory compromise is needed between the negative effects introduced by derivative action and integral action in IOPID controllers, and hence, a more enhanced type of IOPID controller has been evolved with s and $1/s$ being replaced by s^{μ} and $1/s^{\lambda}$ (where, $\mu, \lambda \in [0, 1]$), respectively, in the general IOPID structure [6]. Such a powerful controller with flexible design is the fractional-order PID (FOPID) controller that has been introduced [7, 8] to achieve a perfect balance between relative stability, elimination of high-frequency noise, increased gain, and zero steady-state error, which has gained a lot of significance in the last few decades. FOPID controllers exhibit better step response performance [9] and require less control effort compared to PID controllers. The concept of an FOPID control system is illustrated in **Figure 1**, where $x(t)$ input signal, $e(t)$ error signal, $G(S)$ system or plant transfer function, $y(t)$ output signal, $u(t)$ controller signal, K_p proportional parameter, K_i integral parameter, K_d derivative parameter, λ power of "s" in integral action, μ power of "s" in derivative action. The FOPID controller form is represented mathematically as follows:

$$u(t)_{fc} = K_p \cdot e(t) + K_i \cdot D^{-\lambda} \cdot e(t) + K_d \cdot D^{\mu} \cdot e(t) \tag{1}$$

The continuous transfer function of the $PI^{\lambda}D^{\mu}$ controller is obtained by the Laplace transform as

$$G_c(s) = K_p + K_i \cdot s^{-\lambda} + K_d \cdot s^{\mu} \tag{2}$$

In FOPID, two parameters order of differentiator (λ), and order of integrator (μ) are added along with the three gains of IOPID K_p, K_i, K_d to provide two more degrees of freedom for tuning the controllers so that desired performance such as transient dynamics, stability, and steady state can be easily achieved. The IOPID controller is a subset of the fractions $PI^{\lambda}D^{\mu}$ controller and only the four corner points of the interior square of **Figure 2** can prevail for the IOPID controller. The $PI^{\lambda}D^{\mu}$ controller expands

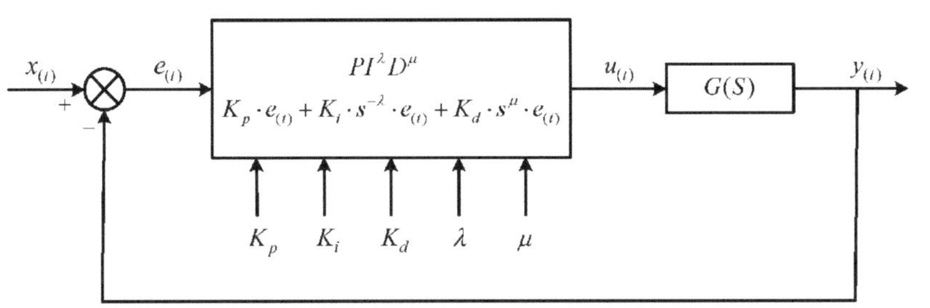

Figure 1.
Fractional-order PID control system.

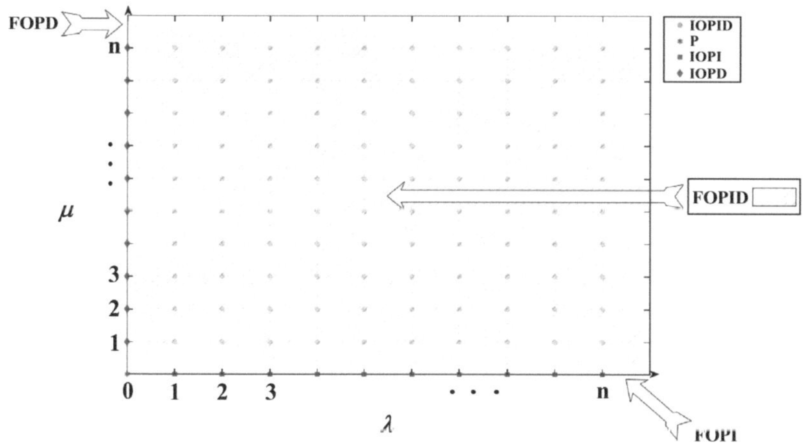

Figure 2.
Various types of PID controllers.

Controller		Fractional-order parameters	
Category	**Type**	λ	μ
PID	P	0	0
	PI	1	0
	PD	0	1
	PID	1	1
$PI^{\lambda}D^{\mu}$	PI^{λ}	$0<\lambda<1$	0
	PD^{μ}	0	$0<\mu<1$
	$PI^{\lambda}D^{\mu}$	$0<\lambda<1$	$0<\mu<1$

Table 1.
Parameters of controllers derived from $PI^{\lambda}D^{\mu}$ controller.

the IOPID controller from point to plane, thus increasing its area of applicability. Indeed, an IOPID controller is obtained when λ and μ are equal to 1. With the FOPID controller, various specifications of the controller can be achieved as given in **Table 1**. The derived expressions of the transfer function, magnitude/phase responses, and integration/differentiation constants for the IOPID and FOPID controllers are given in **Table 2**.

FOPID controller has various advantages over the IOPID controller. Some are as follows: (i) provides supplementary degrees of freedom, which leads to (a) more design flexibility for system design, (b) more precise control of dynamic systems is obtained, (ii) no steady-state error, (iii) more robustness to variations in the gain of the plant, (iv) more robustness to high-frequency noise, (v) better output disturbance rejection, (vi) less sensitive to parametric variations, (vii) it has memory because of fractional integro-differential operations, (viii) attains the property of iso-damping easily, and (ix) performs better and achieves adequate tracking response for higher-order systems, highly nonlinear systems, high time-delay processes, and non-minimum phase systems.

Expression	Integer-order PID controller	Fractional-order PID controller
Transfer function	$C(s) = K_p + \frac{K_i}{s} + K_d \cdot s$	$C(s) = K_p + \frac{K_i}{s^\lambda} + K_d \cdot s^\mu$
Magnitude response	$\lvert C(\omega)\rvert = \sqrt{K_p^2 + \left(\frac{K_i}{\omega} - K_d\omega\right)^2}$	$\lvert C(\omega)\rvert = \left\{K_p^2 + \left(\frac{K_i}{\omega^\lambda}\right)^2 + (K_d\omega^\mu)^2 + \cdots \right.$ $\cdots + 2K_p\left[\left(\frac{K_i}{\omega^\lambda}\right)\cos\left(\frac{\lambda\pi}{2}\right) + K_d\omega^\mu\cos\left(\frac{\mu\pi}{2}\right)\right] +$ $\left.\cdots + 2\left(\frac{K_i}{\omega^\lambda}\right)(K_d\omega^\mu)\cos\left(\frac{(\lambda+\mu)\pi}{2}\right)^{1/2}\right\}$
Phase response	$\angle C(\omega) = -\frac{\pi}{2} + \tan^{-1}\left[\frac{K_d\omega - \left(\frac{K_i}{\omega}\right)}{K_p}\right]$	$\angle C(\omega) = \left\{-\frac{\lambda\pi}{2} + \cdots \right.$ $\left.\cdots + \tan^{-1}\left[\frac{K_p\omega^\lambda\sin\left(\frac{\lambda\pi}{2}\right) + K_d\omega^{(\lambda+\mu)}\sin\left(\frac{(\lambda+\mu)\pi}{2}\right)}{K_i + K_p\omega^\lambda\cos\left(\frac{\lambda\pi}{2}\right) + K_d\omega^{(\lambda+\mu)}\sin\left(\frac{(\lambda+\mu)\pi}{2}\right)}\right]\right\}$
Integration constant	$K_i = \frac{1}{\tau_i}$	$K_i = \left(\frac{1}{\tau_i}\right)^\lambda$
Differentiation constant	$K_d = \tau_d$	$K_d = (\tau_d)^\mu$

Table 2.
Expressions of the transfer function, magnitude and phase responses, and integration and differentiation constants for the IOPID and FOPID controllers.

Nevertheless, the FOPID controller is difficult to tune and realize. Therefore, the pace at which FOPID controllers are embraced in industry has been relatively slow. This chapter intends to analyze the reasons for this trend by examining recent research that outlines the advantages of FOPID controllers, strategies for their implementation, and their various applications in the industrial sector. Furthermore, the specific advantages associated with the utilization of FOPID controllers that arise from frequency-domain analysis are elucidated, accompanied by pertinent conclusions. Moreover, an examination of the potential challenges concerning the industrialization of FOPID controllers is conducted.

2. Literature survey of FOPID controllers

Podlubny [7, 8] has proposed a generalized PD^μ controller and a generalized $PI^\lambda D^\mu$ controller. He also demonstrated the superior performance of the fractional-order controller (FOC) in comparison with the integer-order controller (IOC) when used for the control of FO systems. In study by Podlubny et al. [10], a FO-I^λ controller has been realized within a finite band using the continued fraction expansions (CFE) approximation method. From the existing literature, it is observed that the FOC has been designed to control the IO plant as well as the FO plant. In this connection, two types of design approaches are adopted—one is analytical and the other is optimization-based. In the analytical approach, the continuous-time pure FOC is designed from the frequency-domain design specifications, namely gain crossover frequency, phase margin, flat phase characteristics around gain crossover frequency, etc. On the other hand, the optimization method is based on minimizing the integral square error (ISE), the integral time absolute error (ITAE), or some frequency-domain design specifications. After designing the pure FOCs employing either an analytical approach or an optimization approach, it is approximated either in continuous-time domain or in discrete-time domain. For continuous-time realization of the FOC, the modified Oustaloup approximation method has been used in the study by Xue et al. [11]. The well-known Oustaloup approximation has also been employed in many cases for the same purpose [12, 13]. Discrete-time realization of the FOC can be done either using the indirect rational approximation methods or using the direct rational approximation methods. In the study by Monje et al. [14], the FOPID controller has been approximated first using the Oustaloup method, and subsequently, it is discretized using the Tustin operator. In the study by Folea et al. [15], a FOPD controller is first approximated using the Oustaloup approximation method. Then, it has been discretized with a mixed Euler and Tustin-based transform. In the study by Li et al. [16], a FOPI controller has been approximated using a recursive direct discretization scheme, namely the MRTO method.

The FOPID controller's circuit realization utilizing operational transconductance amplifiers (OTAs) was presented by Suksang et al. [17], where the fractional-order differentiator (FOD) and fractional-order integrator (FOI) stages were created using the suitable RC networks. A FOPID controller was designed by Caponetto et al. [18] using operational amplifiers (OAs) along with an RC network. Oustaloup interpolation was employed to realize the transfer functions of the FOD and FOI stages. Dimeas et al. [19, 20] presented the circuit realization of the PI^{λ} controller for throttle and brake control of an autonomous vehicle (AV) and $I^{\lambda}D^{\mu}$ controller for DC motor speed control using the OTA-C approach. OTA-C-based circuits provided a resistorless realization, enabling the electronically tuning of controller gains through the DC bias current of each OTA. This scheme presented several advantages, including a wider frequency band, lower power dissipation, and suitability for integration. A FOD and an IO integrator with fewer OTAs were used in a modified $I^{\lambda}D^{\mu}$ controller circuit that was published in the study by Kapoulea et al. [21]. This design resulted in a substantial decrease in power consumption. An innovative topology for FOI and FOD was proposed by Kapoulea et al. [22, 23] employing a single extra input second-generation current conveyor (Ex-CCII), along with resistors and capacitors. An AV's throttle and brake control was implemented using this method for the PI^{λ} controller. A PD^{μ} controller for position servo was realized using a single CCII and passive elements, minimizing the active component count [24]. Maurya et al. [25] proposed a $PI^{\lambda}D^{\mu}$ control method for the quad-rotor unmanned aerial vehicles (UAV) position control. The Newton-Euler method is used to formulate a complex model of the quad-rotor. *Via* simulations of trajectory tracking, the efficiency of the planned $PI^{\lambda}D^{\mu}$ method was successfully analyzed. The genetic algorithm (GA) was used to obtain optimal $PI^{\lambda}D^{\mu}$ controller parameters for the desired system specifications. Maurya et al. [26] proposed a FOPID controller designed for higher-order systems. The ITAE is minimized by employing the GA. The GA minimizes the performance metric as well as provides the optimal FOPID controller variables. In another study, Liu [27] proposed an adaptive particle swarm optimization (PSO) algorithm with ITAE for conducting an automatic setting of parameters of FOPID controller. Numerous methods were put forth in this study to alter the cost or fitness objective functions or take into account hybridizing PSO with other optimization algorithms in order to enhance the performance of the classical PSO algorithm. When compared to the differential evolution (DE) algorithm and the normal PSO algorithm, the simulation results reveal that adaptive PSO has the best control over the parameter optimizing result. The FOPID controller parameters have been determined by Karimi et al. [28] using the PSO algorithm, including a new time-domain and frequency-domain performance criterion consisting of various control objectives such as steady-state error, maximum overshoot, rise time, settling time, squared control signal, integral absolute error (IAE), inverse of phase margin, and gain margin for an automatic voltage regulator (AVR) system. Additionally, comparisons were conducted with an IOPID controller, which showed that the proposed FOPID controller could highly improve the system's robustness with respect to model uncertainties. Biswas et al. [29] presented an intelligent optimization method to design the FOPID controllers based on the DE algorithm. A Tustin operator-based CFE approach was used to digitally realize the FO closed-loop transfer function of the designed plant. The superiority of the proposed approach is demonstrated by the performance of the DE-based $PI^{\lambda}D^{\mu}$ controller, particularly for actuating FO plants, when compared to the hierarchical PSO with time-varying acceleration coefficients and binary GA. Both PSO and GA techniques were used by Korkmaz et al. [30] to prove the supremacy of FOPID controller over IOPID. They found FOPID controller advantageous over IOPID controller and also found it more robust to load disturbance. Moreover, PSO was found to be better and faster than GA to get the optimum parameters of the controller. Results proved the supremacy of FOPID controllers and PSO algorithm. Babu et al. [31] implemented FOPID optimization using GA and ant colony optimization (ACO) techniques and also found FOPID to have more robust stability and performance characteristics than PID controller. IOPID and FOPID controllers are designed by Merrikh et al. [32] to control a nonlinear boost converter utilizing the artificial bee colony (ABC) algorithm. Anchored in the simulation results, the authors assert the significance of the *"proposed FOPID controller can capable of improving the start-up response of the boost converter by employing fewer on-off switching actions in comparison to the optimal PID controller"* and emphasize the practical significance of this result. The implementation and analysis of the FOPID controller for speed regulation in a chopper-fed DC motor drive were presented by Rajasekhar et al. [33]. A newly evolved ABC algorithm improved with opposition search techniques was utilized to carry out the optimization task. The performance was compared with the conventional ABC, PSO, and GA methods to demonstrate the effectiveness of the opposition-based ABC. Additionally, Rajasekhar et al. [34] presented a novel hybrid differential ABC algorithm for designing a FOPI speed controller in a permanent magnet synchronous motor drive. The theory and application of a new designing method of the FOPID controller for boost converters were proposed by Tehrani et al. [35]. The method is based on the use of a multi-objective optimization evolutionary algorithm (EA) called strength pareto evolutionary algorithm (SPEA). The simulation and some experimental results proved the superiority of the FOCs over the IOCs. Gao et al. [36] employed a typical FOPID control strategy for the gun control system. To obtain optimal parameters of the controller, a novel Pareto optimal solution-based multi-objective DE algorithm was proposed. Saleh et al. [37] modeled and simulated the brushless DC motor in MATLAB/Simulink for speed control strategy. The FOPID speed controller parameters were optimized by the genetic evolutionary algorithm (GEA) programming module. The simulation results clearly indicate that the proposed FOPID controller results in significant improvements in steady-state error, overshoot, rising time, and settling time. Furthermore, it ensures both flexibility and robust stability when compared to the same system using the IOPID controller. Bingul et al. [38] proposed optimization and design of an IOPID controller and FOPID controller tuned by PSO and ABC algorithms. A non-minimum phase zero and time delay, which are commonly used in process control, were used to test the validity of the controller tuning algorithms in two different systems. The controller tuned by ABC performs better dynamically than controllers tuned by the PSO in terms of time-domain performance criteria such as steady-state error, overshoot, rising time, and settling time. Furthermore, the results of the robustness analysis demonstrated that the ABC-tuned controller parameters are highly robust to internal and external disturbances. Rajesh [39] developed a novel FOPID controller reference model for a nonlinear conical tank system. The PSO method is used to optimize the FOPID controller variables. When compared to IOPID controllers, this comprehensive examination of the obtained data demonstrates that the suggested FOPID controller, which uses the PSO with a reference model, provides smoother control signals, higher stability, and superior qualitative performance. A FOPID controller to minimize certain performance indices such as IAE, ISE, and ITAE has been designed by Kesarkar et al. [40]. The design-construction lead to a high-dimensional, multi-modal, complex optimization problem that can be solved heuristically using an ABC algorithm. A comparison was done between the performances of ABC with a "deterministic" Nelder-Mead simplex algorithm. Goharimanesh et al. [41] proposed an optimized control of type 1 diabetes. It was taken into consideration as a nonlinear model, which was implemented in MATLAB/Simulink. The control method and parameters were tuned by conducting GA. By utilizing FOPID, the controller performance improved considerably, and unlike the IOPID, the blood glucose concentration was maintained in the desired range. To optimize the FOPID controller gains to enhance the ride comfort of heavy vehicles, Gad et al. [42] proposed a multi-objective genetic algorithm (MOGA). The suggested semi-active control algorithm consisted of a system controller that determined the desired damping force using a FOPID controller tuned with a MOGA, as well as a continuous state damper controller that calculated the input voltage to the damper coil. To modify the contact force of the dampers, Zamani et al. [43] devised a FOPID controller utilizing the Cuckoo Search Algorithm (CSA). The ideal option for operating the damper is established based on the results of the fitness function. In comparison with other procedures, this strategy yielded the best results. However, it was designed particularly for seismic excitations and is not appropriate for hybrid energy generation systems. In Ref. [44], the design of an optimal FOPID controller for a full vehicle nonlinear active suspension system (SS) is presented. The FOPID controller's optimal values, in order to minimize the cost function, are tuned using an EA, which present an optimal solution to a multidimensional rough objective function. The results obtained from the full vehicle nonlinear SS using the FOPID controller are compared with those from the corresponding passive SS (system without a controller). The controlled SS has been examined under typical vehicle maneuvers: cornering, sharp braking, and cruising on rough road surfaces. The effectiveness and robustness of the suggested controller have been explicitly demonstrated by the results. A FOPID controller to improve the positioning ability of conducting polymer actuators has been proposed by Itik et al. [45]. The CSA and PSO algorithms were used to obtaining the optimal controller parameters, and a comparison was made between the two algorithms in terms of convergence rate and success of converging to an optimal solution. Furthermore, to evaluate the performance of the FOPID controller, the PID controller was tuned using both algorithms. The results demonstrated that compared to the FOPID controller tuned with PSO, the one tuned with CSA had provided less overshoot, settling, and rising time.

In terms of transient and steady-state response, the PID controller performed somewhat worse than the FOPID controllers. The design of FOPID controllers has been described by Kundu et al. [46], using a newly developed variant of invasive weed optimization (IWO), known as invasive weed optimization with stochastic selection (IWOSS). The FO closed-loop transfer function of the designed plant-controller setup was digitally realized using the Tustin operator-based CFE scheme. Simulation results for some real-life analog plants and comparisons among IWOSS, PSO, and GA were presented to support the claim of superiority of the proposed design technique. An improved ABC algorithm based on the cyclic exchange neighborhood and chaos (CNC-ABC) has been proposed by Zhang et al. [47] to optimize the parameters of the FOPID controller for an AVR system. The performance of the proposed method was evaluated by comparing it with the results of the PSO and GA algorithms. Simulation experiments demonstrated better performance of CNC-ABC-FOPID controller than those of ABC-FOPID controllers, PSO-PID and GA-PID. A delayed 1st-order system is controlled using a FOPID controller [48]. In this paper the authors used GA to optimize the controller response. The Cauchy mutated artificial bee colony (C-ABC) algorithm was used to identify $PI^{\lambda}D^{\mu}$ controller [49]. A novel hybrid control methodology, which is a combination of fuzzy logic and FOPID, was introduced in Ref. [50], and it is applied to an electronic throttle. Fruit fly optimization (FOA) algorithm was used to identify the optimum controller parameters. The authors mentioned that the FO fuzzy PID control gave better results than the IO fuzzy PID control. Das et al. [51] have proposed a FO fuzzy PID controller for a nonlinear and unstable time-delay system. The GA is often used to tune the controller by minimizing several performance measures as well as the control signal. In each case, the system performances and controller activities are compared with those of IOPID controllers, fuzzy PID controllers, and FOPID controllers exposed to various integral performance measures. In maximum cases, simulation results depict that the suggested FO fuzzy PID controller exceeds the other controllers. Hajiloo et al. [52] used a multi-objective Pareto GA to create an optimal fuzzy FOPID controller for FO plants. The objective functions are defined using time and frequency performance measures. When compared to the FOPID controller, the suggested fuzzy FOPID controller outperformed it. Liu et al. [53] have proposed a new tuning method for a variable-order fractional fuzzy PID (VOFFPID) controller for a class of FO and IO plant models. The simulation results depict that VOFFPID controllers outperform FOPID and IOPID controllers in terms of system performance, with the minimum rise time and settling time as well as overshoot, and that the proposed VOFFPID controller could achieve enhanced robustness and flexibility in a variable-order control system. An imperialist competitive algorithm (ICA) was used in Ref. [54] to control the interconnected power systems in three locations. The paper indicated that use of FOPID controller in the system produced a smoother response with reduced oscillations in the steady state. The fractional calculus was introduced in the PSO algorithm to update the velocity of particles and named it the fractional particle swarm optimization-based memetic algorithm (FPSOMA) [55]. The paper used FPSOMA algorithm for the design of a FOPID controller for trajectory control applications. Gases Brownian motion optimization (GBMO) algorithm is used to optimize the response of a hybrid fuzzy FOPID controller for load frequency control in power systems [56]. To control a pump storage unit, an FOPID controller is proposed, and its parameters are optimized using the Caushy and Gaussian mutated gravitational search algorithm (CGGSA) [57]. A novel model reference adaptive control-based FOPID controller was developed for different systems [58] whose parameters are optimized using the moth-flame optimization (MFO) algorithm. To control the movement of robotic arm, FOPID controllers are used, whose parameters are optimized using the colliding bodies optimization (CBO) algorithm [59]. In Ref. [60], a tuning method based on the salp swarm optimization algorithm (SSA) was presented to improve the dynamic response and stability of the FOPID controllers. ITAE was utilized to assess fitness and the optimal result. As a result, by lowering ITAE, ideal FOPID parameters may be attained. If the algorithm becomes stuck in a local solution, the system may incur a significant overshoot and take longer to settle. The design of the FOPID controller based on the sine-cosine algorithm (SCA) was discussed in Ref. [61] and enhanced the performance of the FOPID controller. The chaotic yellow saddle goatfish algorithm (CYSGA) was used by the authors in Ref. [62] to find the optimal parameters for the FOPID controller. In the paper, the authors used a new objective function with a combination of steady-state error, ITAE, settling time, and overshoot. In Ref. [63], the tuning of PID, proportional integral derivative double derivative (PIDD2), and FOPID controllers was accomplished using a combination of two meta-heuristic algorithms: manta ray foraging optimization (MRFO) and simulated annealing (SA). The authors verified that PID controllers are good for set-point tracking, whereas FOPID controllers generate better rise time and settling time values. A multi-objective method based on extremal optimization (MOEO) algorithm was proposed in Ref. [64] and showed that the algorithm produces better tuning parameters than non-dominated sorting genetic algorithm II (NSGA-II) for AVR system. The ant swarm algorithm is hybridized with chaotic maps to optimize FOPID controller for AVR system [65]. The authors showed that the chaotic maps identified better FOPID controller parameters than the previous methods. Lahcene et al. [66] implemented SA for optimizing AVR system by tuning the parameters of FOPID controller. The step responses of the AVR system controlled by SA-FOPID had better performance than those of PSO-FOPID controller, multi-objective liaisons (MOL)-FOPID controller, multi-objective chaotic NSGA-II-FOPID controller, and MOEO-FOPID controller. To tune the parameters of FOPID controller for AVR system, neural networks (NNs) are used in Ref. [67]. The authors showed that NNs are able to optimize controller responses better than the existing optimization algorithms. Juan et al. [68] have proposed a NN-based FOPID controller for a FOLPD pulp consistency system. The FOPID controller was tuned using an radial basis function (RBF) neural network. The simulated results depict that NN-FOPID outperforms IOPID controllers and IOPID controllers based on NNs in terms of time-domain performance specifications and robustness. Hekimoglu et al. [69] proposed chaotic atom search optimization (ChASO) to determine the best FOPID controller parameters. The convergence rate is enhanced since this approach employs logistic map chaotic sequences. It also offers a high level of stability, with minimal overshoot and steady-state deviation. Pradhan et al. [70] developed an Ant Lion Optimization (ALO) algorithm for tuning FOPID controllers by improving the specified performance measures of various systems, like 2nd-order unstable and delay systems, 1st-order delay systems, and higher-order systems. The simulated results illustrate that the suggested ALO algorithm improves the delay system's gain and phase margin, which are normally difficult to find due to the system's non-minimum phase performance. Also, the robustness analysis for the systems under consideration exposes that the suggested technique is more robust to gain variation and produces the best robustness and transient performance characteristics. Moafi et al. [71] proposed an energy management system based on a fuzzy FOPID controller. The fuzzy FOPID controller was found to be superior to the fuzzy logic PID controller and PID controller under various conditions and to have better transient stability after a fault. Yang et al. [72] proposed a new passive FOPID controller for a photovoltaic (PV) grid-connected inverter. Grouped gray wolf optimizer method (GGWO) was used for optimal tuning. The controller achieved maximum power point tracking through a perturbation and observation technique under different weather conditions. For a grid-connected PV system, Yang et al. [73] devised a perturbation observer-based FOPID control mechanism. The controller can estimate the perturbation, which includes multiple nonlinearities, and then successfully correct it. When the Yin-Yang-Pair optimization (YYPO) technique is applied to quickly and efficiently find its optimum control settings, the dynamic performance is significantly enhanced. Muftah et al. [74] designed a PSO fuzzy FOPID controller to reduce overshoot in a double-acting pneumatic actuator's reaction time by employing the PSO algorithm to create a fuzzy FOPID controller. Considering dynamic performance criteria such as time of rise, resting time, and peak overshoot, the PSO fuzzy FOPID controller performs better than the PSO fuzzy PID controller. To improve the stability and transient response, Jumani et al. [75] presented a jaya optimization (JO) algorithm-based FOPID controller. Instead of employing algorithm-specific parameters, this algorithm searches for the best possible solution using just the common governing parameters. Compared to other evolutionary approaches, this reduces the computing time and complexity. Different schemes are used to tune the parameters of the fractional order model reference adaptive controller (FOMRAC), FOPID, and PID controller variables in Ref. [76]. The authors used sequential quadratic programming (SQP), PSO, and GA to identify the controller parameters and showed that the FOPID controller performed better than all other techniques. In order to fine-tune the FOPID controller for automatic generation control (AGC) of hybrid power systems, Irudayaraj et al. [77] applied the atom search optimization (ASO). The suggested controller reduces the settling time and decreases overshoots. The primary concern with the suggested controller is that its performance diminishes quickly in response to changes in climatic conditions. For coordinated management of power consumption in hybrid systems, an AGC based on a FOPID controller is proposed by Khezri et al. The SCA for tackling optimization problems is employed to simultaneously optimize the FOPID controller parameters [78]. Shayeghi et al. [79] looks at the best design and performance assessment of a FOPID controller for load frequency control in an interconnected multi-source power system using the social spider optimization (SSO) method. The results demonstrate that the suggested control strategy can effectively regulate power system frequency deviations even when there are instabilities, non-linearity in the power system structure, and external disturbances.

In recent times, several FOPID controllers were proposed using a multitude of optimization algorithms for DC motors. The algorithms include ABC algorithm [80], JO algorithm [81], and SSA [82]. A FOPID controller parameter tuning method using PSO and constrained PSO to regulate the speed of a DC motor was presented in Ref. [83]. The authors showed that FOPID controllers produced better response and reduced overshoot compared to the IOPID controller. A gray wolf optimization (GWO)-tuned FOPID controller for adjusting DC motor speed was suggested, coupled with rigorous analysis in Ref. [84]. The simulation results indicate that the GWOFOPID method, which uses ITAE as its objective function, achieves less settling and rise times and comparable overshoot compared to existing methods. Hybridization of MRFO and SA techniques incorporated with opposition-based learning (OBL) to find the optimum parameters of FOPID controller for DC motor speed control [85].

The capabilities of the developed algorithm were investigated by comparing it with the original MRFO algorithm and other state-of-the-art algorithms such as GWO, SCA, ASO, and stochastic fractal search (SFS) algorithm. After conducting extensive simulations of the time and frequency domains, as well as robustness and load disturbance rejection analyses, it was found that the proposed method is superior to its counterparts in terms of providing better performance criteria such as less overshoot, less rise and settling times, high gain, and phase margins with high bandwidth. Also, the performance of the developed algorithm has also been demonstrated to be better by using a physical setup. FOPID controller developed in Munagala et al. [86] for DC motors using the Harris Hawks optimization (HHO) algorithm and improved the existing performances. The performance of HHO-FOPID/PID controllers was evaluated using different objective functions (Zwe-Lee Gaing, ITAE, and ITSE) and compared against existing IWO-FOPID/PID, SFS-FOPID/PID, and GWO-FOPID/PID controllers. A novel optimization algorithm is proposed by hybridizing the improved electromagnetism algorithm and genetic algorithm (IEMGA), and it is used for the tuning of FOPID controller for a 2nd-order system [87]. The authors compared the performances of PID controller and FOPID controller and showed FOPID controller produced a better response. A self-organizing migrating algorithm (SOMA) is used to optimize the FOPID controller parameters [88]. In this paper, the authors compared the performance of classical optimization schemes and meta-heuristic algorithms. Interestingly, the results showed that both methods produced similar optimum responses for the system. A real-coded extremal optimization (RCEO) method with multi-non-uniform mutation has been presented by Zeng et al. [89] for the design of FOPID controller. The RCEO is significantly simpler than the reported popular EAs (e.g., chaotic anti swarm (CAS), PSO, and GA) because it has fewer adjustable parameters and just uses selection and mutation operators. Moreover, comprehensive simulation results on both the AVR system and the multivariable control system demonstrated that the RCEO-based FOPID controller proposed in this study is superior to other reported EAs-based FOPID and PID controllers in terms of accuracy and robustness. Verma et al. [90] optimized FOPID controller using GWO. The GWO showed improved optimization in rise time, peak time, and maximum overshoot, but it failed to optimize settling time as compared to those found through other techniques. The design of PID and FOPID controllers to stabilize a ferromagnetic ball in a magnetic levitation system and to control the ball position for tracking a reference signal within an operating region was achieved by Maji et al. [91] through the implementation of PSO, bacterial foraging (BF) optimization, and BF optimized by PSO (BF-PSO) optimization algorithms. The results revealed that BF-PSO provides a superior tuning algorithm compared to the individual BF and PSO algorithms. Further, it proved FOPID works better than PID with BF-PSO tuning approach. Mahmood et al. [92] established a hybrid combination of BF and PSO algorithms that has been done and applied on a FOPID controller for servomotor position dynamics that reduce objective function directly on line. Simulation results demonstrate that the FOPID controller exhibits improved time-domain performance with less value of ISE compared to the IOPID controller; besides, the system with FOPID controller is more flexible and robust than with IOPID controller. Robust stabilization of roll motions has been studied by Faeighi et al. [93]. In order to determine the appropriate values of FOPID controller parameters, chaos-embedded PSO (CPSO) algorithm was used. It concluded that the control system can stabilize the roll motion magnificently in the presence of uncertainties and disturbances.

3. Tuning methods of FOPID controllers

In recent years, various design approaches have been suggested in the literature to improve the FOPID tuning, for instance, using the Ziegler-Nichols (ZN) type method for tuning rules; using gain-phase margin and iso-damping specifications; using IAE and maximum sensitivity; using a flat phase constraint and two other design specifications; using minimizing performance indices; and using a filter with maximum sensitivity specification. Some tuning rules have been successfully implemented in a variety of practical non-integer examples. Many researchers have proposed several methods for tuning FOPID controllers [8]. Valerio et al. [94] proposed a ZN-type empirical rule for tuning FOPID controllers, and fractional M_s constrained integral gain optimization (F-MIGO)-based tuning rules for FOPI controllers have been developed by Chen et al. [95]. However, these methodologies [94, 95] require that the reduced-order models of a higher-order process take only the first-order-plus-dead-time (FOPDT) form, which may not be sufficient to describe the complex dynamic behavior of the higher-order plant. From specified phase margin, gain crossover frequency, and iso-damping/robustness criteria, a tuning methodology for FOPI/FOPD controllers for controlling integer-order systems has been discussed in Refs. [96, 97]. Luo et al. [98, 99] have extended the same tuning strategy for a class of FO models. Monje et al. [14] and Dorcak et al. [88] have proposed an optimization-based frequency-domain tuning method for FOPID controller that incorporates two additional specifications: the complementary sensitivity function and the maximum value of the magnitude of sensitivity, in addition to the specifications presented in Refs. [97–99]. Dominant pole placement tuning [29, 100] and optimal tuning [101–104] based on time-domain integral performance index [105] minimization are two time-domain techniques for FOPID controller tuning. Biswas et al. [29] and Maiti et al. [100] proposed the dominant pole placement tuning, which is applicable exclusively to 2nd-order type systems and fails to deliver satisfactory results for higher-order systems that have several dominant poles and/or zeros. Furthermore, the dominant pole placement tuning [29, 100] often results in unstable responses and yields inferior closed-loop performance for time-delay systems because the Pade approximation of the delay term effectively increases the overall system's order. Cao et al. [101, 102] and Maiti et al. [103] proposed the time-domain optimal tuning method, which often fails to guarantee the close-loop stability of the process along with the FOPID controller. A time-domain tuning method for a FOPID controller utilizing optimization of an integral performance criterion was presented by Deniz et al. [106]. The method is based on Fourier series to obtain the step response of the closed-loop control systems precisely and to accurately obtain the value of error function. Dastjerdi [107] proposed the tuning rules for FOPID controllers using classical loop-shaping tools to locate the initial tuning point applicable for industrial users swiftly. The method proposed in Ref. [108] demonstrates strong performance in linear systems as well as motion systems. Subsequently, to improve set-point tracking and enhance disturbance rejection for FO plants, Li et al. [108] introduced the tuning of the internal model control (IMC) FOPID through the use of maximum sensitivity. The goal of the design is to achieve robustness against process uncertainties and parametric variation. Vinopraba et al. [109] described a single degree of freedom FOPID controller scheme for an IMC-based FO system. Mingjie Li et al. [110] projected a FO system with dead-time (DT) for an IMC-based two degree of freedom (2DOF) FOPID controller. Because a 2DOF control structure keeps the properties of a traditional IMC, it can provide an efficient result if properly tuned. The DT system is estimated by a Taylor approximation, and afterwards several kinds of 2DOF FOPID controllers for three categories of FO systems with DT can be acquired. The simulated outputs validate that the suggested technique is also not easy to develop and tune, but it also improves the system characteristics. Senol et al. [111] presented that a FOPID controller be tuned to improve the performance, stability, and robustness of a 2nd-order delay system. The tuning method focuses on the frequency frame method, which is a rectangular frame encompassing phase and gain margins in the Bode plot and limited by phase and gain crossover frequencies. Padula et al. [112] explained a set of optimal tuning criteria for IOPID and FOPID controllers for integrated and unstable systems. The IAE is used as a performance metric for optimization of servo and regulatory abilities. Alomoush [104] optimized the integral of time multiplied by absolute error, and Lee et al. [87] optimized ISE as the integral performance index to find out the optimal set of controller parameters. Bouafoura et al. [113] have proposed an optimization-based controller tuning method that minimizes matrix norms as the cost functions. Castillo et al. [114] proposed a methodology for tuning FOPI controllers for first-order systems according to frequency-domain specifications and at the same time satisfying several sets of time-domain specifications. For networked control systems, Bhambhani et al. [115] presented a multi-objective optimization-based FOPI controller tuning methodology that concurrently minimizes the closed-loop system's ITAE and maximizes the jitter margin. Therefore, it can be observed that many time-domain integral performance indices have been optimized by various contemporary researchers. The finiteness of the integral performance indices for FO systems for step input and load disturbance excitation is briefly described by Tavazoei [105], and it is necessary to consider this before optimization. Caponetto et al. [116] investigated stabilization of first-order plants with time-delay processes with FOPID controllers. Hamamci et al. [117] addressed a similar stabilization problem using an FOPD/FOPID controller for IO integrating processes, while Hamamci [118] examined the issue for FO integrating processes. Bayat et al. [119] have studied a FOPID controller design for FO systems using the extended root-locus method. A set of tuning rules for IOPID and FOPID controllers have been presented by Padula et al. [120]. The tuning rules were developed based on a FOPDT model of the process and were designed to minimize the IAE while limiting the maximum sensitivity. It was found that using a FO derivative action provides an improvement in performance, whereas using a FO integral action is not advantageous. Ahn et al. [121] developed a new strategy for tuning a FOID controller that meets gain and phase margin requirements, using Bode's ideal transfer function as a reference model for a temperature profile tracking. This analysis yielded a nonlinear equation that establishes a relationship between user-defined gain and phase margins and the FOC. Das et al. [122] did a comparative study on the time and frequency-domain tuning strategies for FOPID controllers to handle higher-order processes and noticed that the frequency technique produces major improvement of robustness (iso-damping), superior abilities of higher frequency noise rejection, lesser control signal value, and thus lowered actuator size, whereas the optimal tuning of the time-domain method is quicker, has less reliability, and has a high risk of failure. Enhancing this work further, Sierociuk et al. [123] presented a tuning algorithm of a variable FOPID controller for switching objective functions. The traditional ZN methods were employed to obtain the time constant parameters of the PID controller, after which the orders of integral and derivative actions were obtained through minimization for different objective functions separately. Mahto et al. [124] introduced a novel FO fuzzy PID controller in which the parameters of the investigated model are tuned using a quasi-oppositional harmony search (QOHS) technique that accelerates convergence. With regard to both linear and nonlinear operating modes, the FOFPID controller outperforms the IOPID and FOPID controllers. Cervera et al. [125] applied quantitative feedback theory (QFT) for the tuning of a FOPID controller. This study analyzed the potential of the FOPID controller to efficiently solve robust control problems and found that FOPID controllers give a very good balance between complexity (number of parameters) and capacity to fit to frequency-domain restrictions in comparison with other fractional structures previously used. Arunachalam et al. [126] presented a robust FOPID controller based on ZN tuning method designed for a cardiac pacemaker to regulate the heart rate. The stable FOPID controller demonstrated superior performance compared to PID controllers with different tuning methods and also outperformed fuzzy logic controllers concerning rise time and percent overshoot. Although the proposed FOPID controller is suitable for rate-adaptive pacing, its design is essentially based on selective tuning processes.

4. FOPID controllers versus high-order integer-order controllers

Higher-order IO transfer function approximations are typically used for real-life implementation of FOPID controllers [6]. This raises the question, *"If IO approximations are used anyway, why not just use high-order IOCs instead of FOPID approximations?"* Although this issue is very important from a modeling perspective, there are surprisingly few publications that address it. A few (almost) relevant papers are reviewed below [127].

In Ref. [128], the author proposes certain rules for choosing the parameters for approximating FOPID controllers using the Oustaloup recursive filter method, potentially assisting in decreasing the order of the resulting controller. Nonetheless, this approach does not solve the *"FOPID versus high-order IOC"* issue that is the topic of this section. In one instance, the authors of Ref. [129] propose the utilization of FO models for "*compressing*" high-order IO models. The suggested method exhibits some constraints (e.g., it is applicable solely to real transfer function poles). In Ref. [130], this approach is expanded. In this work, the authors employ FO models to create compact models acquired through (i) identifying a high-order IO model; (ii) transforming the model into a FO model; and (iii) optimizing the FO model. All of these contributions put forward the fundamental idea that FO models can be used for the compact characterization of high-order systems; however, this fact will not put a stop to the continuous debate about the significance of FOC for control practice. Manifestly, further pertinent research endeavors addressing this issue are required. As a conclusion, we propose the following argumentation [127]:

- Currently, the control community is concentrating on the FOPID controller as an extension of the IOPID controller and is addressing its two additional parameters as "*tuning knobs*." Nevertheless, the implementation of a FOPID controller is based on the use of a high-order IO approximation, and, therefore, all parameters of the resulting approximated system potentially give rise to even more tuning knobs.

- FOPID controllers are commonly not equivalent to high-order IOCs, and we also assert that a near-ideal realization of a FOPID controller can be achieved by implementing a mixed-mode digital-analog system design. In order to implement more complex FO system architectures, this design can use the near-ideal realization of analog FO elements that can be integrated by a digital circuit.

However, effectively managing (e.g., tuning) all the parameters of a high-order IOC is a greater challenge compared to that of a FOPID controller.

5. Industrial applications of FOPID control

A survey paper concerning the industrial use of FOPID controllers was included in Ref. [131]. The main content of the paper consists of the description of various classical control techniques (PID control, backstepping, sliding mode control, MRAC) extended to utilize the FO calculus. A single example of a control problem is given in which a control system for a cement mill is developed. In Ref. [132], the use of FO control of unmanned aerial vehicles (UAVs) and unmanned ground vehicles (UGVs) was investigated. FO control has been applied to the various problems of attitude control, trajectory tracking, path planning, formation control, state estimation, collision avoidance, fault-tolerant control, fault detection, and diagnosis, and has been found to be effective in all these areas; however, the applications are obviously limited when it comes specifically to FOPID type controllers. Nonetheless, the implementation of FOCs is once again highlighted as a significant concern. A comparison of the FOPID and PID controllers for controlling the pressure and flow of a double-shield tunnel boring machine was introduced by Fei et al. [133]. The results demonstrate that employing FOPID controllers for flow and pressure control can effectively reduce fluctuations in flow rate and pressure while also enabling the flow rate and pressure to align with desired set values more quickly. The FOPID controllers for twin rotor systems are being designed by the authors of Ref. [9]. Here, the authors clearly demonstrate that "*... disturbance rejection achieved through the use of the (setpoint weighted FOPID) controller is always more effectively executed in comparison with the designed (setpoint weighted PID), (filtered FOPI) and (filtered PI) controllers ...*" Benlahbib et al. [134] proposed a FOPID controller for the operation of a stand-alone hybrid generating system comprising a wind turbine and a battery bank. The proposed FOPID controller is tested both theoretically and experimentally using MATLAB/Simulink and an experimental testing platform with step changes in wind speed and load impedance. The observed findings reveal improved transient performances in terms of reduced settling time and lower overshoot as well as a minimal steady-state error; nonetheless, the feasibility of integrating the system into the grid should be pursued. A FOC for a wind turbine generator is designed in Ref. [135]. The authors make the following assertions based on simulations: (i) "*... the FOPI controller tracks the input command properly, while the basic IOPI controller fails to handle the nonlinearity arising from the backlash phenomenon.*" and (ii) "*... the FO control system accurately tracks the reference input (under plant parameter variation) ... however, the IO control system becomes unstable (under the same conditions).*" emphasizing the advantages of FOPID controllers once more. Gul et al. [136] proposed a FOPID controller for a pulse-width modulation-based three-phase voltage source inverter in a residential building with integrated PV and wind turbine systems. With fluctuating load circumstances, there is a voltage imbalance concern. Khosravi et al. [137] demonstrated how to manage the frequency of an islanded microgrid using a robust FOPID controller. The FOPID parameters are established by minimizing a set of constraints that assure the system's stability and performance. The main issue with the proposed controller is the time delay.

The reviewed literature provides robust evidence that FO control offers substantial advantages in industrial applications. In order to achieve a seamless transition from theoretical contributions to practice, further research in the realm of industrial applications of FOPID controllers is essential.

5.1 Industrialization of FOPID controllers

According to feedback received from several industrial partners [138], the major problem regarding the adoption of FOPID controllers in industry can be encapsulated in the following question: *"Can the complexities involved in implementation and the associated cost issues of FOPID controllers outweigh the advantages gained from the additional tuning flexibility that FO control provides?"* The industrialization of controllers is much of a challenge nowadays in a competitive world where companies are increasingly investing resources to reduce production costs and enhance the benefits of their products for consumers. As can be seen from the studies in the literature, FOPID controllers have technical advantages over their IO counterparts; however, the production costs of these controllers and the cost-benefit ratio for end users remain areas for further research. Although the FOPID controller in Ref. [9] demonstrates superior technical performance in a two-rotor system, the superior performance does not imply that the expenses associated with developing such FOPID controllers for commercial or industrial use are similar to those of IOPID controllers. In such a case, it is proposed to analyze how to industrialize FOPID controllers while taking financial factors into account. This question is still open and should be considered in upcoming research efforts [127].

The most important step to take toward the industrialization of the FOPID controller is that the FOPID controller must be proven to be at least as reliable as an IOPID controller. It is anticipated that more research will be released, further validating the reliability of FOPID controllers and achieving higher technology readiness levels. This supplementary research will probably provide greater confidence of industrial partners toward FOPID technology, thereby making its integration into industrial more seamless.

6. Conclusions and future trends

This chapter provides an in-depth analysis of pertinent publications in the rapidly growing FO control literature, highlights the milestones, advantages, and disadvantages, and presents future perspectives for the industrialization of FO control. A FOC offers five possibilities for parameter tuning and, consequently, more possibilities for adjusting the roots of the characteristic equation according to the requirements of the imposed performance. Results from research papers demonstrate that transient response specifications of overshoot, settling time, etc., are met by FOCs with lesser control effort and better error performance indices. The robustness of FOC across a wide range of gains is one of its main advantages. The robustness of FOCs against parameter variations makes them an excellent choice for systems, which are difficult to control. The utilization of multi-objective optimization techniques has reduced the complexity of the design of FOCs (but if there are fewer design specifications than the number of controller parameters, the advantages are limited) and makes the tuning of the FO parameters easier. Heuristic optimization algorithms can be used for multi-objective optimization for tuning; however, the random search nature of such algorithms necessitates conducting statistical analysis (such as mean and standard deviation) and performing multiple runs to achieve a dependable solution. Additionally, it is crucial to carefully select cost functions to ensure the most effective results. Many modern-day researchers have focused on the development of novel strategies for tuning FOPID controllers. Most tuning methods utilize a set of design specifications for tuning in both the frequency and time domains. To simplify tuning of FOPID controllers, researchers have proposed a method using reduced IO process models, mainly FOPDT. While using reduced process models facilitates the tuning of FOPID controllers, it also introduces model uncertainties that may lead to an undesirable decline in control performance. Despite all the advantages of flexibility in control, the adoption of FOCs by the industry is still a major concern due to the issues of complexity and cost of implementation. Most of the FOs are approximated by using higher IO transfer functions, which increases the computational resources and cost. This could be rectified by using the analog realization of FO controllers. Developments in the analog IC implementations of various FO elements can be used to realize various FOCs at lower cost.

6.1 Future trends

- Further research may be required to specify simpler parameter tuning knobs for FOPID controllers to be accepted in the industry.

- Plug-in-type FOPID controllers with inbuilt functionalities may be developed for industrial applications, which are currently still dominated by IOPID controllers.

- With the advances in artificial intelligence and fuzzy logic, better optimization techniques can be used to tune the FOPID controllers.

In spite of the comprehensive reference list presented here, consisting of mostly research papers from the last two decades, the area of FO calculus in PID controller methods remains a subject that requires further exploration. In particular, many problems related to the application of such methods on real processes need to be solved. This chapter details how FO calculus has made its mark on PID controller methods with multiple useful consequences.

Author details

Yunus Emre Yokuş[1,2*], Umut Engin Ayten[2] and Aslıhan Kartcı[2]

1 Department of Electronics and Communication Technology, Army NCO Vocational HE School, National Defence University, Balıkesir, Turkey

2 Department of Electronics and Communications Engineering, Yildiz Technical University, İstanbul, Turkey

*Address all correspondence to: emre.yokus@std.yildiz.edu.tr; yemreyokus@gmail.com

IntechOpen

© 2024 The Author(s). Licensee IntechOpen. This chapter is distributed under the terms of the Creative Commons Attribution License (http://creativecommons.org/licenses/by/4.0), which permits unrestricted use, distribution, and reproduction in any medium, provided the original work is properly cited.

References

[1] O'Dwyer A. PID control: The early years. In: Control in the IT Sector Seminar. Ireland: Cork Institute of Technology; 2005

[2] Åström KJ et al. PID Controllers: Theory, Design and Tuning. Research Triangle Park, NC: Instrument Society of America; 1995

[3] O'Dwyer A. Handbook of PI and PID Controller Tuning Rules. London: Imperial College Press; 2009

[4] Åström KJ et al. Towards intelligent PID control. Automatica. 1992;**28**:1-9

[5] Xue D et al. Linear Feedback Control: Analysis and Design with MATLAB. Society for Industrial and Applied Mathematics; 2007

[6] Monje C et al. Fractional-Order Systems and Controls: Fundamentals and Applications. Springer; 2010

[7] Podlubny I. Fractional-Order Systems and Fractional-Order Controllers. Slovakia: Institute of Experimental Physics, SAS; 1994

[8] Podlubny I. Fractional-Order Systems and PIλDμ-Controllers. Vol. 44. IEEE Transactions on Automatic Control; 1999. pp. 208-214

[9] Azarmi R et al. Analytical design of fractional order PID controllers based on the fractional set-point weighted structure: Case study in twin rotor helicopter. Mechatronics. 2015;**31**:222-233

[10] Podlubny I et al. Analogue realizations of fractional-order controllers. Nonlinear Dynamics. 2002; **29**:281-296

[11] Xue D et al. Fractional order PID control of a DC-motor with elastic shaft: A case study. Proceedings of the American Control Conference. 2006: 3182-3187

[12] Monje C et al. Proposals for fractional PIλDμ tuning. In: Proceedings of the first IFAC symposium on fractional differentiation and its applications. Vol. 38. 2004. pp. 369-381

[13] Valério D et al. Tuning-rules for fractional PID controllers. IFAC Proceedings Volumes. 2006;**39**:28-33

[14] Monje C et al. Tuning and auto-tuning of fractional order controllers for industry applications. Control Engineering Practice. 2008;**16**(7): 798-812

[15] Folea S et al. Discrete-time implementation and experimental validation of a fractional order PD controller for vibration suppression in airplane wings. Acta Polytechnica Hungarica. 2017;**14**(1):191-206

[16] Li W, et al. Design of fractional-order PIα controller with two modes. In: IEEE International Power Electronics and Motion Control Conference. Vol. 3. IEEE. 2006. pp. 1-5

[17] Suksang T et al. Implement the fractional-order, half integrator and differentiator on the OTA base PIλDμ controller circuit. In: 9th International Conference on Electrical Engineering/ Electronics, Computer, Telecommunications and Information Technology. 2012. pp. 1-4

[18] Caponetto R et al. Integrated technology fractional order proportional-integral-derivative design. Journal of Vibration and Control. 2014; **20**(7):1066-1075

[19] Dimeas I et al. OTA-C realization of PIλ brake and throttle controllers for autonomous vehicles. In: IEEE European Conference on Circuit Theory and Design. IEEE; 2017. pp. 1-4

[20] Dimeas I et al. New analog implementation technique for fractional-order controller: A DC motor control. AEU - International Journal of Electronics and Communications. 2017; **78**:192-200

[21] Kapoulea S et al. Simple design of fractional-order DC motor controller. In: 6th International Conference on Control, Decision and Information Technologies. 2019. pp. 867-871

[22] Kapoulea S et al. Single active element implementation of fractional-order differentiators and integrators. AEU - International Journal of Electronics and Communications. 2018; **97**:6-15

[23] Kapoulea S et al. Single active element fractional-order PIλ controller for vehicle throttle and brake control. In: 8th International Conference on Modern Circuits and Systems Technologies. 2019. pp. 1-4

[24] Kapoulea S et al. CCII based realization of fractional-order PD controller for a position servo. In: 42nd International Conference on Telecommunications and Signal Processing. 2019. pp. 102-105

[25] Maurya HL et al. Trajectory tracking of quad-rotor UAV using fractional order PIλDμ controller. In: Computational Intelligence: Theories, Applications and Future Directions-Volume I. Singapore; 2019. pp. 171-186

[26] Maurya AK et al. Tuning of fractional order PID controller for higher order process based on ITAE minimization. In: Annual IEEE India Conference. 2015. pp. 1-5

[27] Liu X. Optimization design on fractional order pid controller based on adaptive particle swarm optimization algorithm. Nonlinear Dynamics. 2016;**84** (1):379-386

[28] Karimi-Ghartemani M et al. An optimal fractional order controller for an AVR system using particle swarm optimization algorithm. In: Large Engineering Systems Conference on Power Engineering. 2007. pp. 244-249

[29] Biswas A et al. Design of fractional order PIλDμ controllers with an improved differential evolution. Engineering Applications of Artificial Intelligence. 2009;**22**:343-350

[30] Korkmaz M et al. Performance comparison of fractional controllers and artificial intelligence techniques. International Journal of Machine Learning. 2013;**3**:229-232

[31] Babu AGS et al. Implementation of fractional order PID controller for an AVR system using GA and ACO optimization techniques. IFAC. 2016;**49**: 456-461

[32] Merrikh-Bayat F et al. Comparing the Performance of Optimal PID and Optimal Fractional-Order PID Controllers Applied to the Nonlinear Boost Converter. arXiv preprint arXiv: 1312.7517, 2013

[33] Rajasekhar A et al. Fractional order PID controller design for speed control of chopper fed DC motor drive using artificial bee colony algorithm. In: World Congress on Nature and Biologically Inspired Computing. 2013. pp. 259-266

[34] Rajasekhar A et al. A hybrid differential artificial bee colony

algorithm based tuning of fractional order controller for permanent magnet synchronous motor drive. International Journal of Machine Learning and Cybernetics. 2014;**5**:327-337

[35] Tehrani K et al. Design of Fractional Order PID Controller for Boost Converter Based on Multi-Objective Optimization. In: Proceedings of 14th International Power Electronics and Motion Control Conference; 2010. pp. T3-T179

[36] Gao Q et al. Multiobjective optimization design of a fractional order PID controller for a gun control system. The Scientific World Journal. 2013;(1): 907256

[37] Saleh A et al. Speed control of brushless DC motor based on fractional order PID controller. International Journal of Computer Applications. 2014;**95**(4)

[38] Bingul Z et al. Comparison of PID and FOPID controllers tuned by PSO and ABC algorithms for unstable and integrating systems with time delay. Optimal Control Applications and Methods. 2018;**39**(4):1431-1450

[39] Rajesh R. Optimal tuning of FOPID controller based on PSO algorithm with reference model for a single conical tank system. SN. Applied Sciences. 2019;**1**(7):758

[40] Kesarkar A et al. Tuning of optimal fractional-order PID controller using an artificial bee colony algorithm. Systems Science & Control Engineering. 2015;**3** (1):99-105

[41] Goharimanesh M et al. Fractional order PID controller for diabetes patients. Journal of Computational Applied Mechanics. 2015;**46**(1):69-76

[42] Gad S et al. Multi-objective genetic algorithm fractional-order PID controller for semi-active magnetorheologically damped seat suspension. Journal of Vibration and Control. 2017;**23**(8): 1248-1266

[43] Zamani A et al. Fractional order PID control design for semi-active control of smart base-isolated structures: A multi-objective cuckoo search approach. ISA Transactions. . 2017;**67**:222-232

[44] Aldair AA et al. Design of fractional order controller based on evolutionary algorithm for a full vehicle nonlinear active suspension systems. International Journal of Control and Automation. 2010;**3**:33-46

[45] Itik M et al. Fractional order control of conducting polymer artificial muscles. Expert Systems With Applications. 2015; **42**(21):8212-8220

[46] Kundu D et al. Designing fractional-order PIλDμ controller using a modified invasive weed optimization algortihm. In: World Congress on Nature & Biologically Inspired Computing. 2009. pp. 1315-1320

[47] Zhang D et al. Optimum design of fractional order PID controller for an AVR system using an improved artificial bee colony algorithm. Acta Automatica Sinica. 2014;**40**(5):973-979

[48] Zhang Y et al. Fractional-order PID controller tuning based on genetic algorithm. In: International Conference on Business Management and Electronic Information. Vol. 3. 2011. pp. 764-767

[49] Rajasekhar A et al. Fractional-Order PIλDμ Controller Design Using a Modified Artificial Bee Colony Algorithm. In: Swarm, Evolutionary, and Memetic Computing: Second International Conference. 2011. pp. 670-678

[50] Sheng W et al. Fruit fly optimization algorithm based fractional order fuzzy-PID controller for electronic throttle. Nonlinear Dynamics. 2013;**73**:611-619

[51] Das S et al. A novel fractional order fuzzy PID controller and its optimal time domain tuning based on integral performance indices. Engineering Applications of Artificial Intelligence. 2012;**25**(2):430-442

[52] Hajiloo A et al. Fuzzy fractional-order PID controller design using multi-objective optimization. In: Joint IFSA World Congress and NAFIPS Annual Meeting. 2013. pp. 1445-1450

[53] Liu L et al. Variable-order fuzzy fractional PID controller. ISA Transactions. 2015;**55**:227-233

[54] Taher S et al. Fractional order PID controller design for LFC in electric power systems using imperialist competitive algorithm. Ain Shams Engineering Journal. 2014;**5**(1):121-135

[55] Mousavi Y et al. A memetic algorithm applied to trajectory control by tuning of fractional order proportional-integral-derivative controllers. Applied Soft Computing. 2015;**36**:599-617

[56] Zamani A et al. Design of a fractional order PID controller using GBMO algorithm for load–frequency control with governor saturation consideration. ISA Transactions. 2016;**64**:56-66

[57] Li C et al. Design of a fractional-order PID controller for a pumped storage unit using a gravitational search algorithm based on the Cauchy and Gaussian mutation. Information Sciences. 2017;**396**:162-181

[58] Bourouba B et al. Moth-flame optimisation algorithm-based fractional order PIλDμ controller with MRAC tuning configuration. International Journal of Systems, Control and Communications. 2018;**9**:148-171

[59] Mohammadi, Asl R et al. Optimal fractional order PID for a robotic manipulator using colliding bodies design. Soft Computing. 2018;**22**: 4647-4659

[60] Khan IA et al. Salp swarm optimization algorithm-based fractional order PID controller for dynamic response and stability enhancement of an automatic voltage regulator system. Electronics. 2019;**8**(12):1472

[61] Bhookya J et al. Optimal FOPID/PID controller parameters tuning for the AVR system based on sine–cosine-algorithm. Evolutionary Intelligence. 2019;**12**:725-733

[62] Micev M et al. Fractional order PID controller design for an AVR system using chaotic yellow saddle goatfish algorithm. Mathematics. 2020;**8**(7):1182

[63] Micev M et al. Optimal design of automatic voltage regulation controller using hybrid simulated annealing–manta ray foraging optimization algorithm. Ain Shams Engineering Journal. 2021;**12**(1): 641-657

[64] Zeng G et al. Design of fractional order PID controller for automatic regulator voltage system based on multi-objective extremal optimization. Neurocomputing. 2015;**160**:173-184

[65] Tang Y et al. Optimum design of fractional order PIλDμ controller for AVR system using chaotic ant swarm. Expert Systems With Applications. 2012; **39**(8):6887-6896

[66] Lahcene R et al. Optimal tuning of fractional order pid controller for avr

system using simulated annealing optimization algorithm. In: 5th International Conference on Electrical Engineering-Boumerdes. 2017. pp. 1-6

[67] Bhullar A et al. Design of fopid controller for optimizing avr system using neural network algorithm. In: 17th India Council International Conference. 2020. pp. 1-7

[68] Shan WJ et al. A neural network fractional order PID controller for FOLPD process. In: 35th Chinese Control Conference. 2016. pp. 10459-10463

[69] Hekimoğlu B. Optimal tuning of fractional order PID controller for DC motor speed control via chaotic atom search optimization algorithm. IEEE Access. 2019;7:38100-38114

[70] Pradhan R et al. Optimal fractional order PID controller design using ant lion optimizer. Ain Shams Engineering Journal. 2020;**11**(2):281-291

[71] Moafi M et al. Energy management system based on fuzzy fractional order PID controller for transient stability improvement in microgrids with energy storage. International Transactions on Electrical Energy Systems. 2016;**26**: 2087-2106

[72] Yang B et al. Grouped grey wolf optimizer for maximum power point tracking of doubly-fed induction generator based wind turbine. Energy Conversion and Management. 2017;**133**: 427-443

[73] Yang B et al. Perturbation observer based fractional-order PID control of photovoltaics inverters for solar energy harvesting via yin-yang-pair optimization. Energy Conversion and Management. 2018;**171**:170-187

[74] Muftah MN et al. Modeling and fuzzy FOPID controller tuned by PSO for pneumatic positioning system. Energies. 2022;**15.10**:3757

[75] Jumani T et al. Jaya optimization algorithm for transient response and stability enhancement of a fractional-order PID based automatic voltage regulator system. Alexandria Engineering Journal. 2020;**59**(4): 2429-2440

[76] Ortiz-Quisbert M et al. Optimal fractional order adaptive controllers for AVR applications. Electrical Engineering. 2018;**100**:267-283

[77] Irudayaraj A et al. A Matignon's theorem based stability analysis of hybrid power system for automatic load frequency control using atom search optimized FOPID controller. IEEE Access. 2020;**8**:168751-168772

[78] Khezri R et al. Coordination of heat pumps, electric vehicles and AGC for efficient LFC in a smart hybrid power system via SCA-based optimized FOPID controllers. Energies. 2018;**11**(2):420

[79] Shayeghi H et al. Multi-source power system FOPID based load frequency control with high-penetration of distributed generations. In: 21st Conference on Electrical Power Distribution Networks Conference. IEEE; 2016. pp. 131-136

[80] Mishra A et al. Speed control of dc motor using artificial bee colony optimization technique. In: International Conference on Control, Automation, Robotics and Embedded Systems. 2013. pp. 1-6

[81] Achanta RK et al. DC motor speed control using PID controller tuned by jaya optimization algorithm. In: International Conference on Power, Control, Signals and Instrumentation Engineering (ICPCSI). 2017. pp. 983-987

[82] Hekimoğlu B et al. Speed control of DC motor using PID controller tuned by salp swarm algorithm. In: Proc. IENSC. 2018. pp. 1878-1889

[83] Jain R et al. Tuning of fractional order PID controller using particle swarm optimization technique for DC motor speed control. In: IEEE 1st International Conference on Power Electronics, Intelligent Control and Energy Systems. 2016. pp. 1-4

[84] Agarwal J et al. Analysis of grey wolf optimizer based fractional order PID controller in speed control of DC motor. Microsystem Technologies. 2018;**24**: 4997-5006

[85] Ekinci S et al. Optimal FOPID speed control of DC motor via opposition-based hybrid manta ray foraging optimization and simulated annealing algorithm. Arabian Journal for Science and Engineering. 2021;**46**:1395-1409

[86] Munagala VK et al. Design of fractional-order PID/PID controller for speed control of DC motor using Harris hawks optimization. In: Intelligent Algorithms for Analysis and Control of Dynamical Systems. Singapore: Springer; 2021

[87] Lee C et al. Fractional-order PID controller optimization via improved electromagnetism-like algorithm. Expert Systems With Applications. 2010;**37.12**: 8871-8878

[88] Dorcak L et al. Design of the fractional-order PIλDμ controllers based on the optimization with self-organizing migrating algorithm. Acta Montanistica Slovaca. 2007;**12**:285-293

[89] Zeng G et al. A real-coded extremal optimization method with multi-non-uniform mutation for the design of fractional order PID controllers. Information Technology and Control. 2016;**45**(4):358-375

[90] Verma SK et al. Optimization of fractional order PID controller using grey wolf optimizer. Journal of Control, Automation and Electrical Systems. 2017;**28**:314-322

[91] Maji L et al. Design of PID and FOPID controllers based on bacterial foraging and particle swarm optimization for magnetic levitation system. In: Proceedings of Indian Control Conference Indian Institute of Technology Madras. 2015. pp. 463-468

[92] Mahmood AK et al. Design of fractional order PID controller based on hybrid bacterial for aging-particle swarm optimization. International Journal of Engineering and Innovative Technology. 2016;**5**:1-6

[93] Faieghi M et al. Design of fractional-order PID for ship roll motion control using chaos embedded PSO algorithm. In: 2nd International Conference on Control, Instrumentation and Automation. 2011. pp. 606-610

[94] Valério D et al. Tuning of fractional PID controllers with Ziegler–Nichols-type rules. Signal Processing. 2006;**86** (10):2771-2784

[95] Chen Y et al. Practical tuning rule development for fractional order proportional and integral controllers. Journal of Computational and Nonlinear Dynamics. 2008;**3**(2):021403

[96] Monje C et al. On fractional PIλ controllers: Some tuning rules for robustness to plant uncertainties. Nonlinear Dynamics. 2004;**38**:369-381

[97] Li H et al. A fractional order proportional and derivative (FOPD) motion controller: Tuning rule and

experiments. IEEE Transactions on Control Systems Technology. 2009;**18**(2):516-520

[98] Luo Y et al. Fractional order [proportional derivative] controller for a class of fractional order systems. Automatica. 2009;**45**(10):2446-2450

[99] Luo Y et al. Tuning fractional order proportional integral controllers for fractional order systems. Journal of Process Control. 2010;**20**(7):823-831

[100] Maiti D et al. Design of a fractional-order self-tuning regulator using optimization algorithms. In: 11th International Conference on Computer and Information Technology. 2008. pp. 470-475

[101] Cao JY et al. Design of fractional order controllers based on particle swarm optimization. In: 1st IEEE Conference on Industrial Electronics and Applications. 2006. pp. 1-6

[102] Cao J et al. Optimization of fractional order PID controllers based on genetic algorithms. International Conference on Machine Learning and Cybernetics. 2005;**9**:5686-5689

[103] Maiti D et al. Tuning PID and PIλDδ controllers using the integral time absolute error criterion. In: 2008 4th International Conference on Information and Automation for Sustainability. 2008. pp. 457-462

[104] Alomoush M. Load frequency control and automatic generation control using fractional-order controllers. Electrical Engineering. 2010;**91**(7): 357-368

[105] Tavazoei MS. Notes on integral performance indices in fractional-order control systems. Journal of Process Control. 2010;**20**(3):285-291

[106] Deniz FN et al. Tuning of fractional order PID controllers based on integral performance criteria using Fourier series method. IFAC. 2017;**50**:8561-8566

[107] Dastjerdi A et al. Tuning guidelines for fractional order PID controllers: Rules of thumb. Mechatronics. 2018;**56**: 26-36

[108] Li D et al. Maximum sensitivity based fractional IMC–PID controller design for non-integer order system with time delay. Journal of Process Control. 2015;**31**:17-29

[109] Vinopraba T et al. IMC based fractional order PID controller. In: International Conference on Industrial Technology. 2011. pp. 71-76

[110] Li M et al. Two-degree-of-freedom fractional order-PID controllers design for fractional order processes with dead-time. ISA Transactions. 2016;**61**:147-154

[111] Şenol B et al. Fractional order proportional derivative control for time delay plant of the second order: The frequency frame. Journal of Franklin Institute. 2020;**357**(12):7944-7961

[112] Padula F et al. Optimal tuning rules for proportional integral derivative and fractional order proportional integral derivative controllers for integral and unstable processes. IET Control Theory & Applications. 2012;**6**(6):776-786

[113] Bouafoura MK et al. PIλDμ controller design for integer and fractional plants using piecewise orthogonal functions. Communications in Nonlinear Science and Numerical Simulation. 2010;**15**:1267-1278

[114] Castillo FJ et al. Design of a class of fractional controllers from frequency specifications with guaranteed time domain behavior. Computers &

Mathematics with Applications. 2010;**59**: 1656-1666

[115] Bhambhani V et al. Hardware-in-the-loop experimental study on a fractional order networked control system testbed. Communications in Nonlinear Science and Numerical Simulation. 2010;**15**(9):2486-2496

[116] Caponetto R et al. New results on the synthesis of FO-PID controllers. Communications in Nonlinear Science and Numerical Simulation. 2010;**15**: 997-1007

[117] Hamamci S et al. Calculation of all stabilizing fractional-order PD controllers for integrating time delay systems. Computers & Mathematics with Applications. 2010;**59**(5):1621-1629

[118] Hamamci S. An algorithm for stabilization of fractional-order time delay systems using fractional-order PID controllers. IEEE Transactions on Automatic Control. 2007;**52**(10):1964-1969

[119] Merrikh-Bayat F et al. Method for designing PIλDμ stabilisers for minimum-phase fractional-order systems. IET Control Theory & Applications. 2010;**4**(1):61-70

[120] Padula F et al. Tuning rules for optimal PID and fractional-order PID controllers. Journal of Process Control. 2011;**21**(1):69-81

[121] Ahn HS et al. Fractional-order integral and derivative controller for temperature profile tracking. Sadhana. 2009;**34**:833-850

[122] Das S et al. On the selection of tuning methodology of FOPID controllers for the control of higher order processes. ISA Transactions. 2011; **50**(3):376-388

[123] Sierociuk D, Macias M. Comparison of variable fractional order PID controller for different types of variable order derivatives. In: Proceedings of the 14th International Carpathian Control Conference. 2013. pp. 334-339

[124] Mahto T et al. Fractional order fuzzy PID controller for wind energy-based hybrid power system using quasi-oppositional harmony search algorithm. IET Generation, Transmission & Distribution. 2017;**11**(13):3299-3309

[125] Cervera J et al. Tuning of fractional PID controllers by using QFT. In: IECON 32nd Annual Conference on IEEE Industrial Electronics. 2006. pp. 5402-5407

[126] Arunachalam SP et al. Intelligent fractional-order PID (FOPID) heart rate controller for cardiac pacemaker. In: Healthcare Innovation Point-Of-Care Technologies Conference. 2016. pp. 105-108

[127] Tepljakov A et al. Towards industrialization of FOPID controllers: A survey on milestones of fractional-order control and pathways for future developments. IEEE Access. 2021;**9**: 21016-21042. DOI: 10.1109/ ACCESS.2021.3055117

[128] Merrikh-Bayat F. Rules for selecting the parameters of Oustaloup recursive approximation for the simulation of linear feedback systems containing PID controller. Communications in Nonlinear Science and Numerical Simulation. 2012;**17**(4):1852-1861

[129] Mansouri R et al. Approximation of high order integer systems by fractional order reduced-parameters models. Mathematical and Computer Modelling. 2010;**51**(1-2):53-62

[130] Barbé K et al. Fractional models for modeling complex linear systems under

poor frequency resolution measurements. Digital Signal Processing. 2013;**23**(4):1084-1093

[131] Efe MO. Fractional order systems in industrial automation—A survey. IEEE Transactions on Industrial Informatics. 2011;7:582-591

[132] Cajo R et al. A survey on fractional order control techniques for unmanned aerial and ground vehicles. IEEE Access. 2019;7:66864-66878

[133] Fei L et al. Fractional-order PID control of hydraulic thrust system for tunneling boring machine. In: International Conference on Intelligent Robotics and Applications. Berlin, Heidelberg: Springer; 2013. pp. 470-480

[134] Benlahbib B, et al. Fractional Order PID Controller for DC Link Voltage Regulation in Hybrid System Including Wind Turbine-and Battery Packs-Experimental Validation. International Journal of Power Electronics. 2019;**10**(3): 289-313

[135] Ghasemi S et al. Application of fractional calculus theory to robust controller design for wind turbine generators. IEEE Transactions on Energy Conversion. 2014;**29**(3):780-787

[136] Gül O et al. Application of fractional-order voltage controller in building-integrated photovoltaic and wind turbine system. Measurement and Control. 2019;**52**(7-8):1145-1158

[137] Khosravi S et al. Robust control of islanded microgrid frequency using fractional-order PID. Iranian Journal of Science and Technology, Transactions of Electrical Engineering. 2020;**44**(3): 1207-1220

[138] Vansovits V et al. Identification of industrial water boiler for model predictive control of district heat plant. In: 3th Biennial Baltic Electronics Conference. 2012. pp. 315-318